An Introduction to Quantum Mechanics

W. Blaine Dowler

Preface to the Second Edition

The first edition of this book was released in 9 weekly installments as the first "Summer School" series on http://www.bureau42.com in 2010. This second edition corrects some typographical errors, clarifies some points that resulted in most of the questions, and adjusting the language for the differences between a weekly series and a single volume release.

Contents

Preface to the Second Edition	iii

1 Classical Thinking: Why Does It Fail? — 1
  1.1 Some Preliminary Words 1
  1.2 Classical Pictures 2
      1.2.1 Sphere Problems 2
      1.2.2 Interaction Problems 3
      1.2.3 Nucleus Mass Problems 3
      1.2.4 Information Transmission Problems 4
      1.2.5 Particle Creation Problems 5

2 Curiouser and Curiouser — 7
  2.1 Unanswered Questions 7
  2.2 Energy and Mass 7
  2.3 Mass of a Nucleus: $2 + 2 = 3.9$? 8
      2.3.1 Mass and Inertia 8
      2.3.2 Not So Elementary, My Dear Watson 9
  2.4 Radioactive Decay 9
      2.4.1 Muons and Taus and Neutrinos, Oh My! 10
      2.4.2 Decay Timelines 11

3 Enter Heisenberg, Exit Common Sense — 13
  3.1 Unanswered Questions 13
  3.2 Determinism 14
  3.3 Limitations on Experimental Observations 14
  3.4 The Heisenberg Uncertainty Principle and Measurements . . . 15
  3.5 The Heisenberg Uncertainty Principle and Existence 16
      3.5.1 Philosophical and Theological Interpretations 17

4 Don't Underestimate the Power of Virtual Particle Exchange — 19
  4.1 Unanswered Questions 19
  4.2 Information Exchange 20
  4.3 Conservation of Energy 20
  4.4 The Other Forces 22

4.5	The Strong Nuclear Force	23
4.6	A New Question	24

5 Let There Be Quantized Electromagnetic Radiative Energy 25
- 5.1 Unanswered Questions . . . 25
- 5.2 Some Words About Electric Current . . . 25
- 5.3 Some Words About Waves and Light . . . 26
  - 5.3.1 Shape . . . 26
  - 5.3.2 Speed . . . 26
  - 5.3.3 Wavelength . . . 27
  - 5.3.4 Frequency . . . 27
  - 5.3.5 Intensity . . . 27
- 5.4 The "Neener Neener" Spot . . . 27
- 5.5 The Blackbody Radiation Problem . . . 28
- 5.6 The Photoelectric Effect . . . 29
- 5.7 Quantization of Light . . . 30
  - 5.7.1 Reconciliation Problem . . . 31
  - 5.7.2 Work Function Implications . . . 31

6 Quanta, Quanta Everywhere 33
- 6.1 Unanswered Questions . . . 33
- 6.2 Diffraction . . . 33
- 6.3 Particle or Wave? . . . 35
  - 6.3.1 None of the Above . . . 35
- 6.4 Heisenberg and Existence . . . 35
- 6.5 Inverse Square Laws . . . 36
  - 6.5.1 Classical View . . . 37
  - 6.5.2 The Quantum View . . . 38

7 Down the Rabbit Hole 41
- 7.1 Unanswered Questions . . . 41
- 7.2 Welcome To Wonderland . . . 41
- 7.3 The Rabbit Hole Without the Hole . . . 42
- 7.4 Through the Looking Glass. Or Not. . . . 42
- 7.5 A Window In The Quantum World . . . 43
- 7.6 Electron Orbits . . . 44

8 One and One and One is Three 47
- 8.1 Unanswered Questions . . . 47
- 8.2 Volume of Matter . . . 47
- 8.3 Molecule Formation . . . 48
- 8.4 Electricity and Magnetism . . . 48
- 8.5 Nature's Smallest Bar Magnets . . . 49
- 8.6 The Pauli Exclusion Principle . . . 50
- 8.7 Insulators and Covalent Bonds . . . 50
- 8.8 Insulators and Ionic Bonds . . . 51

	8.9	Conductors	53
		8.9.1 Electrical Resistivity	53
		8.9.2 Photoelectric Conductors and Semiconductors	54

9 Like a Record, Baby — **55**
- 9.1 Unanswered Question ... 55
- 9.2 Angular Momentum ... 55
- 9.3 Spin ... 56
- 9.4 Fermions and Bosons ... 57
- 9.5 Spin and Behaviour ... 58
- 9.6 Conclusion ... 59

Chapter 1

Classical Thinking: Why Does It Fail?

1.1 Some Preliminary Words

One consequence of Einstein's relativity causes serious problems for most classical pictures of the subatomic world. Exactly how this consequence comes about is best covered in a session on relativity, and therefore will not be covered here. All we need is the consequence itself so that we can apply it: **Nothing, not even information, can travel faster than the speed of light.**

There is also one mathematical rule that comes into play quickly: **It is forbidden to divide by zero.** Dividing by smaller and smaller numbers leads to bigger and bigger ratios. (i.e. if you divide by 2, you get a bigger answer than if you divide by 3. Divide by $\frac{1}{2}$, and it gets bigger yet. Divide by $\frac{1}{1000}$ and it is like you just multiplied by 1000. The closer to 0 your divisor gets, the bigger your answer. If you divide by zero, you have just made the answer infinitely large.)[1]

It's also important to state that this first chapter is not about answering questions. Rather, it is about finding the right questions to ask to help us realize that we need to have some sort of bizarre looking theory to explain the seemingly mundane world we inhabit. The answers come later, often followed immediately by even more questions.

---

[1] Dividing by fractions is often hard to picture, frequently due to the verbiage used to establish division in elementary school. If we have 6 objects, and we divide by 3, we are often told to "divide them into three groups" or "divide them into groups of three." Though not wrong, this verbiage makes it unclear what we are doing when we divide by fractions. Think of "divide 6 by 3" as "these 6 objects represent 3 groups; how many are in a single group?" instead. We still naturally arrive at 2 objects per group. However, when dividing by fractions, "6 divided by $\frac{1}{2}$" then becomes "these 6 objects represent half a group," and grasping that there are 12 items in a single complete group becomes much easier, and far more natural.

1.2 Classical Pictures

The classical picture of an atom presented in most public school systems is based on the solar system. Protons and neutrons are, somehow, stuck together in the middle like a sun, called a nucleus and perfectly spherical electrons orbit around this centre like planets through the electromagnetic forces. It's clean, it's familiar, it's easy to picture, and it's completely wrong.

1.2.1 Sphere Problems

Imagine that electrons, protons, and neutrons are, in fact, small spheres. In that model, one could take a handful of electrically neutral and indivisible particles[2] and line them up in a row, as with the beads in a Newton's cradle. Much like the Newton's cradle, you could smack one end and the particle on the other would move. With solid, structureless spheres, that leads to a conceptual problem.

Now imagine you had ten such particles in a row. They would all be identical, including their diameters. Now, let us examine the information transmitted in this process. When you push on the first particle in the line, the information about that push is immediately transmitted to the next particle it is contact with. (A structureless particle is an incompressible particle. Push it on the left, and the right side moves instantaneously.) How quickly is this information transmitted?

The speed of anything is the distance covered divided by the time elapsed. When you push the particle on one end of our line, information about this push is transmitted ten diameters down. So, the information travels a distance of 10 diameters. However, this information travels instantaneously. That means 0 time elapsed. The speed would be ten diameters divided by zero, and division by zero is forbidden. The speed would be infinitely high, and that's a whole lot faster than the speed of light.[3]

If nothing can travel faster than light, then we cannot use this picture of particles. Two options present themselves. In one possibility, the particles we are dealing would have to have some sort of internal structure. This works well enough for a store-bought Newton's cradle, but not for quantum mechanics: at some point, we need to boil things down to the basic building blocks of matter, and that means particles without structure. The other option is to get the physical distance travelled by the information down to 0 as well. Due to this, all indivisible particles, called elementary particles, would have to have zero volume. Only then does instantaneous information transfer become consistent with relativity.

That poses a new problem: if all matter is made up of particles that have no volume, then why does matter take up space at all? Why doesn't it all collapse to a point? How can particles bind together so consistently that we can define a

---

[2]To most people, this means neutrons. These people would be wrong, but that's a topic for the next chapter.

[3]The Newton's Cradle you can buy in a store doesn't have this problem because the balls are compressible, and because there is a time delay between hitting one end and the ball responding on the opposite end. The balls aren't very compressible, and the time delay is very small, but both effects are enough to get information transmission under the light speed limit.

1.2. CLASSICAL PICTURES

specific density for a specific material, keep particles from falling together on each other, and also keep them from getting so far apart that the material collapses? Furthermore, why do we experience any kind of contact forces? Why can we feel a table instead of putting our hands through the table entirely? How do particles with no volume interact at all?

1.2.2 Interaction Problems

Orbits imply interactions. Planets are held in place through interactions involving gravity. When dealing with protons, neutrons, and electrons, gravity is not even remotely strong enough to hold the batch together. There is only one other force in nature that we observe at the macroscopic level, and that's the electromagnetic force. This seems to work for establishing orbits, as electrons and protons would be attracted to each other. Pick the right set of distance scales, and you would think this would work for building atoms out of the nucleus and the electrons.

There is a major and fundamental problem with this picture, too. What the heck holds the nucleus together? neutrons have no electric charge, so there seems to be no way to hold them in place. Protons would actually repel each other. As they would also be packed in tightly, the repulsion between protons should be much stronger than the attraction to electrons. The nucleus should, by classical understanding, tear itself asunder long before complex molecules could form. Clearly, complex molecules such as DNA not only exist but can remain stable for long periods of time, so there is a flaw in this model. We have to have at least one other force in play, and this force must greatly exceed the strength of the electromagnetic force trying to pull the nucleus apart. This leads to yet another problem: if there is at least one other force in nature that is significantly stronger than the electromagnetic force, then why haven't we seen it in the macroscopic world?

We have another problem with this model. Why is every nucleus made from protons and neutrons only? Why don't we ever have electrons in a nucleus as well, or instead of protons? Why can't an atom have electrons and neutrons in its nucleus and protons in orbit?

1.2.3 Nucleus Mass Problems

In 1932, James Chadwick discovered the neutron. He discovered that two nuclei with the same number of protons could still have different masses, and realized that a nucleus must contain more than one type of particle, one of which was electrically neutral. Classical thinking dictates that all particles of a given type are identical, and experiments bear that out: when studied in isolation, all protons have identical properties to all other protons, all neutrons are identical to all other neutrons, electrons to electrons, and so forth. Thus, if you measure the mass of the proton, and the mass of the neutron, you can then predict and calculate the masses of all combinations of protons and neutrons. (i.e. a nucleus composed of two protons and two neutrons would have a mass identical to the combined masses of two protons and two neutrons.) It's a simple idea, which follows from our natural instincts, and which one would expect from identical, indivisible particles combining in different ways.

Too bad it's wrong.

Experiments have shown that the total mass of a nucleus does not match the combined masses of its component parts. Now, one could propose some sort of glue that is holding the particles together, which would increase the combined mass. This would make sense, except the mass of a stable nucleus is invariably less than that of its component particles. Either the glue has a negative mass, or something entirely different is going on. To make matters worse, nuclei that have more mass than the particles they are made out of have a nasty habit of releasing energy, not mass, and then finding themselves with a lower mass as a result.

That leaves us with two problems: why does energy transmission change a particle's mass, and what is gluing these particles together to form nuclei?

1.2.4 Information Transmission Problems

We know of ways to transmit information without wires. Radio waves broadcast music and cellular traffic constantly. Wi-fi networks keep us connected to the Internet throughout homes and coffee shops worldwide. However, the methods we use all involve the input of energy, usually electrical power. That energy must come from somewhere; after all, classical physics dictates that energy is always conserved, meaning the sum total of all energy in the universe does not change. It may change form,[4] but the energy is always around here somewhere. The same is true every time information is transmitted in all of its forms.

If we have an electron orbiting a nucleus, then that electron "knows" of an opposing electrical charge in the nucleus. In other words, information about that charge has been received. In order to manage that, energy needs to be transmitted away from one or the other. Where does that energy come from? How much energy does each particle have to transmit? How does the transmitted energy get replenished? It cannot balance in a direct exchange between the two particles; if that happened, that could only mean that every transmitted piece of energy is exactly balanced by an incoming piece of information. This, in turn, would imply that every transmitted piece of information reaches a destination. That is only guaranteed if the information is only transmitted toward particles ready to receive that information. In that case, then the transmitting particle would already "know" where to find the receiving particles. This creates circular logic: in order to transmit information between two particles while conserving energy, information about the two particles must already be known to both particles.

Nonetheless, information *does* get exchanged between particles, and yet violations of energy conservation have never been observed in an experiment. This seems to be a logical inconsistency that will need to be sorted out.

---

[4] One form energy can take is mass.

1.2. CLASSICAL PICTURES

1.2.5 Particle Creation Problems

The final problem comes not from the observations of the very small particles in quantum mechanics, but in the observations of the very large cosmological entities. Early in the 20th Century, it was discovered that the Universe is expanding outwards in all directions. This implies that it was packed in together far more tightly in the past. The question is, how tightly? In a world of absolutes, where particles are immutable and indivisible, the particles also become invincible. Every massive particle that exists now has always existed and always will exist. This means that the entire Universe would have once been packed into a single, small space with such density that gravity could not be ignored. That much mass in one place could not have catalysed a Big Bang to create the Universe. Thus, we could not have had all of the particles that exist today around at the time. So, what was there? How did it spark? How did all of these particles get created? The last question can be answered in the realm of quantum mechanics. Any scientific studies of the remaining questions belong in cosmology.

CHAPTER 1. CLASSICAL THINKING: WHY DOES IT FAIL?

Chapter 2

Curiouser and Curiouser

2.1 Unanswered Questions

Our discussions in the previous chapter revealed a few problems with classical thinking. To summarize:

1. The basic building blocks of matter, called elementary particles, must all have zero volume. What, then, prevents them from piling in so closely together that the matter they form does not have zero volume?

2. How does a nucleus hold itself together, if the electromagnetic force trying to push the protons apart is so much stronger than gravity, and gravity is the only attractive force we know of that applies?

3. Why aren't electrons ever found in a nucleus?

4. Why is the mass of a nucleus different from (and often lower than) the total mass of its component particles?

5. Why does the mass of a nucleus change when it emits energy, even though that energy has no mass?

6. How does information about particle positions get exchanged between particles? That would seem to violate energy conservation.

7. How are new particles created?

2.2 Energy and Mass

The fifth question on the above list is actually the easiest to answer, not because the answer is intuitive in any way, but because the behaviour is consistent, reproducible, and involves relatively few variables.

Before: we have a nucleus of a given mass.

During: the nucleus spontaneously discharges energy.

After: the nucleus left has the same number of protons, same number of neutrons, and less mass.

The conclusion seems inescapable: it is possible to convert mass into energy. In fact, some more detailed experiments have revealed that the amount of mass "lost" is very closely related to the amount of energy discharged. Specifically, if the mass of the nucleus is reduced by mass m, and if the energy discharged is E, then the relationship between the two is given by the equation $E = mc^2$, where $c = 299,792,458$ m/s $\approx 186,000$ mi/s, which is the speed of light in a vacuum. This equation is most closely associated with Einstein's relativity, because Einstein reached the same conclusion years before quantum physics reached that point, though he approached it from an entirely different direction.

2.3 Mass of a Nucleus: $2 + 2 = 3.9$?

As discussed in chapter one, a nucleus made of a given number of protons and neutrons generally has a mass less than the total masses of the protons and neutrons it is made of. In fact, every stable nucleus (i.e. any nucleus that does not spontaneously fall apart) is guaranteed to have a mass less than the total masses of the parts it is made of. How can this be? Why is the mass not only different, but less than its components? The first step toward finding the solution to this issue is by examining how we are measuring the mass of a nucleus in the first place.

We have yet to develop the kitchen scale that can detect the presence of a single nucleus. The traditional means of determining mass (put it on a scale, find the weight[1] and convert) is simply not an option for objects this small and hard to count. Instead, we have to study the way nuclei move and deduce the mass from this process. Most commonly, physicists would strip a single electron off a set of atoms to make sure they had an overall electric charge,[2] and then launch the atoms through a magnetic field. When a charged particle moves through a magnetic field, it moves in a circular arc. After measuring the radius of the arc produced with a known charge and known magnetic field, the "mass" of the nucleus can then be measured.

The word "mass" in the previous paragraph appears in quotations for a reason. Newton made a mistake.[3] When studying an object in motion, we do not measure the mass of that object.

2.3.1 Mass and Inertia

When Newton first proposed his laws of motion, he made a distinction between two quantities that was ignored almost immediately.

---

[1] Remember, mass and weight are different quantities. Mass is the amount of stuff something is made of, while weight is the force of gravity acting on that much mass. You will still have mass in a weightless environment, and the fastest way to lose weight is to move to a planet with weaker forces of gravity.

[2] In this case, the electric charge is that of a single, unbalanced proton.

[3] In Newton's defence, he couldn't have possibly known any better at the time.

2.4. RADIOACTIVE DECAY

- *Mass*: Mass is the amount of matter an object is made of.
- *Inertia*: Inertia is an object's resistance to changes in motion.

By all experiments available to Newton at the time he formed his theories, mass and inertia appeared to be one and the same. They are not.

We have already seen that energy and mass are closely connected. It stands to reason that, if energy is closely tied to mass, and mass is closely tied to inertia, then energy and inertia would also have a connection. That is, in fact, the case.

What we have been referring to up to this point as "mass" is truly inertia. The mass of a nucleus is not different from the mass of the particles it is composed of. Rather, the inertia, which is a combination of both mass and the energy required to glue that mass together, is different from the inertia of the particles it is composed of.

The first step to the solution has been taken: we realized that the common vernacular includes invalid assumptions dating back centuries.[4] In fact, the terminology is so ingrained that it is still common to refer to inertia as "mass" and the actual mass as "rest mass," meaning the "mass" we measure for a particle when it is not moving. These common conventions will be held for the rest of this book: "mass" will refer to inertia, and "rest mass" will refer to mass.

2.3.2 Not So Elementary, My Dear Watson

We can now explain why we have different values for the mass of a nucleus and the sum of the masses of its components. What we cannot explain is why the mass is reduced when we put a nucleus together.[5]

By what we've seen so far, we would expect the mass of a nucleus to be greater than the total mass of the particles we make it from. It takes energy to stick two particles together, so we should have the mass of those two particles combined, plus the energy (called *binding energy*) that holds them together.

Again, we find the solution is logical but not intuitive. The simplest explanation is this: protons and neutrons are not elementary, indivisible particles. Rather, they are also composed of other constituent particles.[6] This allows the mass of a combination of these particles to be less than the mass of the particles on their own. Through whatever mechanism they use to bind themselves together, it must take less energy to hold a proton's component particles together when it is in a nucleus than it does in isolation. The same is true for neutrons.

2.4 Radioactive Decay

We have already mentioned that some nuclei are stable, while others are not. The stable ones, when left undisturbed, do not change their fundamental nature over

---

[4]This is not the last time this is going to happen through this book. Or the last time on this page, for that matter.
[5]The difference between the mass of a nucleus and the sum of the masses of its component particles is called the *mass defect*.
[6]Specifically, each proton and neutron is composed primarily of three quarks.

time. This is not true of all nuclei.

Some nuclei are unstable. Given time, they will transform into other nuclei, generally emitting other particles in the process. One common example is an isotope of carbon known as carbon-14. Isotopes are numbered by the grand total number of protons and neutrons in the nucleus; as all carbon nuclei have 6 protons, this nucleus also has 8 neutrons. Left alone for a sufficiently long amount of time, this nucleus will "decay," transforming into a nucleus of nitrogen-14, emitting other particles in the process. As with all natural decay processes, the total mass of the ejected particles is less than the mass of the original carbon-14 nucleus. This is the process scientists use for carbon dating, which they use to determine the age of archaeological artifacts. (More on this later.)

The process of radioactive decay is better understood today than it was a century ago. The conservation of energy can be applied to explain why the mass always reduces in natural, spontaneous processes: the total amount of energy in your "ingredients" when you start limits the total mass and energy of the particles that result.

2.4.1 Muons and Taus and Neutrinos, Oh My!

Early experimenters took the opposite approach to decay; if you can add energy to a system, then perhaps you can create more particles. This was tried, by accelerating a particle with electric charge to a high kinetic energy, and then slamming it head-on[7] into another particle. Reactions were mixed. Along with the elation of finding out that the theory worked, and new particles could, indeed be created in this fashion, came the general confusion that resulted by creating particles never before identified by man.

One of the first of these particles to be created was the muon, named for the Greek letter mu (μ) that was arbitrarily assigned to it. In almost all respects, it appeared to be identical to the electron. It has considerably greater mass[8] and is unstable; given time, it will decay, leaving behind an electron and, when first observed, some massively confused researchers. Physicists have long held to the notions of energy and momentum conservation. Both conservation laws are logical and intuitive, and were consistent with centuries of observation. The decay of the muon was the first time these conservation laws were seriously questioned. When the muon decayed, the only observable output was the electron. Let's break down what that means.

Let's say we have a stationary muon; it has no kinetic energy whatsoever.[9] After some time, it decays. The only particle we observe coming out of the decay is an electron, which has significantly lower mass. Thus, if we conserve energy, the electron must be produced in motion; the kinetic energy the electron gains

---

[7] This is merely a figure of speech. To the best of our knowledge, particles with zero volume do not have heads.

[8] The mass of the muon is over 200 times greater than the mass of the electron.

[9] Kinetic energy is the energy of motion. If a particle is stationary, kinetic energy is zero. The faster a particle moves, the more kinetic energy it has. The more massive a particle is, the more kinetic energy it has.

2.4. RADIOACTIVE DECAY

must match the difference in masses times the speed of light squared, by $E = mc^2$. This leads to another problem, however. Momentum[10] would not be conserved; the muon was at rest, but the electron is not. If the electron comes out stationary, then momentum would be conserved, but energy would not. Another observation seems, at first, to make things even worse: the electron ejected by the decay does not, in fact, account for all of the energy of the original muon. The electron's total energy is invariably less than the muon's.

We have a few options to reconcile these issues. One is to throw out one of the conservation laws. They are, after all, arbitrary human rules that have stuck around only because they have worked so incredibly well for centuries. Another is to propose yet another new particle, and a rather exotic one at that. If there were another particle or two leaving the collision, then these particles could reconcile all of the difficulties. Energy and momentum could both be conserved, as these new particles would carry the balance. Moreover, the electron would then be required to have less energy than the muon did, which is consistent with experiment. The great remaining question then is this: why don't the other particles[11] show up in our detectors?

The detectors at the time could only detect particles directly through two forces: electromagnetic forces, and gravity. If the particles in question had no electric charge, then they wouldn't be detectable that way. (Neutrons also show no electric charge.) If their masses were also very small, possibly zero, then we couldn't detect them by gravity. However, if neutrinos exhibit neither of these forces, then a new question arises: what force(s) do they experience? Every measurable interaction in nature is driven by some kind of force. We will need more than two forces to explain neutrinos completely.

As it turns out, energy and momentum are conserved in muon decay, as well as all other decay processes. There are, in fact, other particles in play, called neutrinos. Similarly, there is another electron-like particle, called the tau, which is about 3,490 times more massive than the electron, and which may spontaneously decay into neutrinos and either muons or electrons (among other, less common options.)

2.4.2 Decay Timelines

Radioactive decay is the most common example of the natural creation of particles. This is, then, a reasonable starting point to examine the process. One of the first questions we can ask is "what triggers radioactive decay to produce particles?" We have found that unstable particles can decay when they experience collisions with other particles, or are otherwise given more energy. In the case of individual particles, rather than whole nuclei, becoming a part of a system such as a nucleus

---

[10] Momentum is the product of mass and velocity, or speed. Bullets hurt because they have high momentum due to their high speed, not because of their relatively small mass. Similarly, if a fast moving bicycle hits a slow moving train, bet on the bike taking more damage, as the train's significantly higher mass gives it far more momentum.

[11] Careful studies of these decay processes dictate that not one, but two such exotic particles must be produced in muon decay.

can actually increase the stability of a particle.[12] What happens when the particle is isolated, and left to itself?

This is where things start to get truly strange. Left to itself, an unstable particle or nucleus will decay. There is even a defined rate of decay, called the "half life" of the particle. For example, carbon-14 has a half life of approximately 5,730 years. In other words, if you were to take a million carbon-14 nuclei and leave them alone for 5,730 years, then you'd be left with about half a million carbon-14 nuclei and half a million nitrogen-14 nuclei when you returned. If you left them alone for another 5,730 years, you'd return to approximately a quarter of a million carbon-14 nuclei (and three quarters of a million nitrogen-14.) For each 5,730 years you leave the sample, you'd halve the remaining particle count again. In other words, this is not linear: if you leave for two half-lives, you don't get down to zero, you get down to a quarter of the original.

This has a few implications worth noting.

- The rate of decay does not depend on the original particle count.
- The exact decay of a single particle appears to be a random process.

The decay of carbon-14 also has one property worth noting: the total mass of the output particles (nitrogen-14 nucleus, an electron, and an antineutrino) is less than the mass of the carbon-14 nucleus.

We can predict to some accuracy when a given portion of a sample will decay. We cannot, however, predict when any specific particle will decay. If we had a single carbon-14 nucleus to begin with instead of a million, we couldn't predict exactly when it would decay. There's a 50% chance it would still be carbon after 5,730 years, a 25% chance it would still be carbon after 11,460 years, a 12.5% chance it would still be carbon after 17,190 years, and so forth, but we can never say "this nucleus will decay in exactly four hours and eighteen minutes." If the process were not random, we wouldn't have a definable half life. Either the decay would be triggered by some unrecognized part of the experimental apparatus, in which case two different labs would have two different half lives for the same object, or the decay would be based on some innate "timer" in each particle, meaning we would have a linear decay; if half the sample decayed in 5,730 years, then it would all decay in 11,460 years.

This is the mechanism behind carbon dating; all carbon in nature, including that in living tissue, contains a certain proportion of carbon-14. Living things have mechanisms which will automatically correct errors in proteins and cells if a carbon atom decays into nitrogen. Dead things don't do this as effectively. If radiation detectors determine that half the carbon-14 in an artifact has decayed into nitrogen, then they know the artifact is about 5,730 years old. If they find a different relative proportion, they are working with an artifact or corpse of a different age.

---

[12]The most common example of this would be the neutron. Though quite stable as a part of most nuclei, a free neutron, left to itself, will eventually decay, transforming into a proton, an electron, and an antineutrino.

Chapter 3

Enter Heisenberg, Exit Common Sense

3.1 Unanswered Questions

The questions raised to date which still have not been answered are as follows:

1. The basic building blocks of matter, called elementary particles, must all have zero volume. What, then, prevents them from piling in so closely together that the matter they form does not have zero volume?

2. How does a nucleus hold itself together, if the electromagnetic force trying to push the protons apart is so much stronger than gravity, and gravity is the only attractive force we know of that applies?

3. Why aren't electrons ever found in a nucleus?

4. How does information about particle positions get exchanged between particles? That would seem to violate energy conservation.

5. How are new particles created? Answer currently incomplete.

6. Can the physical laws of the Universe truly allow a random process?

7. What force(s) do neutrinos experience?

At the end of the previous chapter, we learned that radioactive decay is a random process, which takes place at seemingly arbitrary times. This opens up a whole range of possibilities: if there is one random process in nature, then it stands to reason that there would be others.

3.2 Determinism

The first concept this challenges is determinism. Determinism is the belief that everything that happens in nature is absolute, and that a sufficiently advanced scientific model paired with sufficiently complete experimental observations could be used to accurately predict all future events. This has long been an assumption of scientists, but radioactive decay makes it appear to be incorrect. Granted, radioactive decay is not *completely* random,[1] but the inability to accurately predict the moment of decay for any single nucleus is disturbing to many scientists.

To press forward, let us re-examine two of our unanswered questions from the first two chapters:

1. How does information about particle positions get exchanged between particles? That would seem to violate energy conservation.

2. How are new particles created? (The answer to this one is, at the moment, incomplete.)

These two questions can actually be cleared up if we can accept uncertainty in the universe we live in.

3.3 Limitations on Experimental Observations

To make any measurement of a system, one must interact with that system. In doing so, you alter that system. Let's look at a circuit as an example.

There are three main quantities that can be measured in any electric circuit: voltage, current and resistance. Resistance is purely a result of the materials the circuit has been constructed from. Materials with high resistance make it difficult for current to pass through, and often get warm in the process. (Toasters, light bulbs, stoves and ovens are essentially based on this principle.) Current is, essentially, a measure of how many electrons are moving through the circuit at a time. Voltage, finally, is a measure of how much potential energy is available to make the electrons move, or, if you prefer, it a measure of how strongly the electrons are pushed through the circuit. These three variables are related in such a way that you can't change one of the three without changing at least one of the other two.[2]

Electric circuits also form an example of experimental limitations. To measure the current in a circuit, one must put the measurement tool in the circuit along the path the electrons travel. This allows the experimenter to confidently "count" all of the electrons in motion and determine the current. However, this also means increasing the total resistance of the circuit by the resistance of your measurement tool, making it impossible to measure the natural voltage of the circuit.[3] Similarly, to measure the voltage, we must put our measurement tool in parallel in the circuit,

---

[1] We can define a meaningful half-life for the decay rate of any given isotope.

[2] Mathematically, the relationship can be written as $V = IR$, where V is voltage, I is current, and R is resistance.

[3] Well made equipment can keep these distortions at very low levels, to the point where it's well below the threshold of the tools to measure, but all equipment has some impact.

meaning the circuit must be altered to fork into the voltage meter before forking back into a single path. Although this measures the voltage well, the current in either half of the fork alone would not be representative of the current in the entire circuit. If you were to try to measure both current and voltage simultaneously, you would exaggerate both effects.

In short, it is impossible to measure both voltage and current in a circuit with arbitrary accuracy. They can be measured with high accuracy, but there will be imperfections in the combined measurements that can never be entirely eliminated.

3.4 The Heisenberg Uncertainty Principle and Measurements

It is not difficult to imagine that similar limitations as that on measuring voltage and current as described above would exist in other situations, as well. When you get down to the quantum mechanical level, in which distances and subjects are so incredibly small, it is not hard to imagine that there would be more significant limitations of this type.

Any time we take a measurement, we alter the subject of that measurement. In the case of subatomic particles, the effect is significant. As we've already deduced, elementary particles must have zero volume. There's not a microscope in the world that can magnify zero into something greater than zero, so we have to get creative if we want to determine where a particle is. We have to force it to interact with something else that has a known position and calculate where it is. In doing so, the interaction changes the energy of that particle, making it difficult to measure that particle's momentum with any accuracy. Similarly, if one attempts to measure the momentum of a particle, then one disturbs the position of the particle, limiting the accuracy of that measurement.

Werner Heisenberg[4] was the first to formulate a mathematical relationship that determines exactly how limited our measurements of related variables can get. The uncertainty principle he derived still bears his name.

The logic for the principle is pretty straightforward. If measuring the same two quantities in two different orders alters the results, then we can define that uncertainty in terms of the difference between these two orders. We look at the results if we measure position (x) first and momentum (p) second, and multiply them together. (For example, let's pretend this gave us measurements of 50 and 20 respectively, in some appropriate set of units, multiplying out to 1000.) Then, we look at the results if we measure momentum first and position second, and multiply those together. (For example, measuring 21 and 48 respectively, in the same units as our first pair, whatever they may be. These multiply out to 1008.) If we subtract

---

[4]Unrelated trivia: Werner Heisenberg was the scientist in charge of the Nazi atomic bomb project in World War II. At one point, the Nazi project was more advanced than the Allied project, but was abandoned when a test bomb failed to detonate. It was later shown that it failed because Heisenberg made a mistake in his calculations relating to the amount of radioactive fuel required, and the bomb only had 10% of the uranium required for proper detonation. It is still unclear whether or not this computational error was accidental.

these, we will then have some idea of the degree of our uncertainties. The difference between the two measurements is at least 8 (in appropriate units), so the average uncertainty in one pair of measurements is at least 4 (in the same units.) If our equipment is precise enough that we can measure position within 2 units, then our uncertainty in momentum must be at least $4 \div 2 = 2$ units. If our equipment can measure position within 1 unit, then our uncertainty in momentum must be at least $4 \div 1 = 4$ units. At this point, the principle is logical and easy to accept. There is one other aspect of the principle that tends to throw our instincts out of whack, however.

This is not just a limitation on our ability to measure particles, this is a limitation on the *existence* of subatomic particles. The reasons for this are, unfortunately, best left for later chapters.[5]

3.5 The Heisenberg Uncertainty Principle and Existence

We have already seen that radioactive decay involved a random element subject to well defined constraints. (i.e. any individual unstable particle may or may not decay at any given time, but the probability of the decay is defined well enough to produce a measurable half life.) That phenomena demands a level of random, uncertain behaviour crop up in the scientific theory itself. The Heisenberg Uncertainty Principle is what governs that behaviour. This explains why there is random behaviour, why there are limits on how random that behaviour can get, and why we don't notice this behaviour during our day to day life.

Because the uncertainties in quantities are defined in pairs, there is always an indeterminate "wiggle room" for each individual measurement. The particles in question cannot be described in absolutes. They exist in an indeterminate state. That is why we can't predict exactly when an unstable particle will decay; the current state of any given particle is in flux. The flux is limited by the Heisenberg principle, and the limits of the Heisenberg principle are so tight that they are unnoticeable on normal distance scales. This seems to contradict one idea of science: we have always assumed that we can take measurements as precisely as we like (given proper equipment) and that we can then predict the future of a system with absolute certainty. We can't. This is not a limitation on modern technology, nor is it something which merely requires more research. This is the way our world works, like it or not. So, some might ask, what is the point of continuing if we can't make absolute predictions? Well, although we cannot measure within this range of random behaviour, we can measure the probabilities of the outcomes of this behaviour with all the accuracy we require. It is with this mindset that we can happily continue our research.

---

[5]The unfortunate part about trying to build a model of the Universe one block at a time is that the Universe has already been built and exists. All of the building blocks in the model already exist, and sometimes we hit phenomena in combinations that cannot be properly explained in a linear order. Be patient, and we *will* get there.

3.5. THE HEISENBERG UNCERTAINTY PRINCIPLE AND EXISTENCE

3.5.1 Philosophical and Theological Interpretations

The Heisenberg Uncertainty Principle is still a subject of debate. There are three different schools of thought that seem to prevail:

- Some people believe that events taking place within these limits is entirely random.

- Some people believe that there is nothing truly random going on, but that there are "hidden variables" within these limits that we can never measure, but which govern the results of these questions. This allows the Universe to remain a place of absolute rules, even if we can never access all of the information we need to act on all of those rules.

- Some people believe that this "uncertain realm" is the area in which the deity of choice steers the course of reality. Again, we'll never be able to directly measure what's going on in this realm, but the Universe is no longer random.

Due to the nature of the Heisenberg Uncertainty Principle, it is impossible for science to distinguish between these three viewpoints. The author's preference for the first interpretation will almost undoubtedly come through as a bias in the verbiage chosen, but this preference is a completely aesthetic choice. Each reader is encouraged to choose among these viewpoints, or to develop a new one, based on his or her own aesthetic preferences. In short, science and religion are, indeed, compatible, and neither science nor religion will ever be able to describe existence in its totality. No warranty is expressed or implied. Your mileage may vary.

Chapter 4

Don't Underestimate the Power of Virtual Particle Exchange

4.1 Unanswered Questions

Our collection of questions is impressive indeed:

1. The basic building blocks of matter, called elementary particles, must all have zero volume. What, then, prevents them from piling in so closely together that the matter they form does not have zero volume?

2. How does a nucleus hold itself together, if the electromagnetic force trying to push the protons apart is so much stronger than gravity, and gravity is the only attractive force we know of that applies?

3. Why aren't electrons ever found in a nucleus?

4. How does information about particle positions get exchanged between particles? That would seem to violate energy conservation.

5. How are new particles created? Answer currently incomplete.

6. What force(s) do neutrinos experience?

7. How do we know the Heisenberg Uncertainty Principle applies to particle existence, and not merely measurement?

The answers to many of these questions depend on a surprising aspect of the Heisenberg Uncertainty Principle.

4.2 Information Exchange

Recall from chapter one that the biggest issue with information transmission is formed in terms of energy conservation. To quote that lesson:

> If we have an electron orbiting a nucleus, then that electron "knows" of an opposing electrical charge in the nucleus. In other words, information about that charge has been received. In order to manage that, energy needs to be transmitted away from one or the other. Where does that energy come from? How much energy does each particle have to transmit? How does the transmitted energy get replenished? It cannot balance in a direct exchange between the two particles; if that happened, that could only mean that every transmitted piece of energy is exactly balanced by an incoming piece of information. This, in turn, would imply that every transmitted piece of information reaches a destination. That is only guaranteed if the information is only transmitted toward particles ready to receive that information. In that case, then the transmitting particle would already "know" where to find the receiving particles. This creates circular logic: in order to transmit information between two particles while conserving energy, information about the two particles must already be known to both particles.
>
> Nonetheless, information does get exchanged between particles, and yet violations of energy conservation have never been observed in an experiment. This seems to be a logical inconsistency that will need to be sorted out.

The answer to this dilemma turns out to be stranger than anticipated. You may notice some very particular phrasing in the last paragraph: "... violations of energy conservation have never been observed in an experiment." As it turns out, violations of energy conservation occur constantly, but we'll never observe them.

4.3 Conservation of Energy

As discussed in the previous lesson, the Heisenberg Uncertainty Principle is not strictly a limitation on measurement, but it is an actual property of existence. The Universe is truly "uncertain" about details within the limits of this principle. In many ways, the "it's not cheating if you don't get caught" philosophy is ingrained in nature itself. Conservation laws can be bent (or even broken) provided the process is over and done with quickly enough to be unobservable.

Just as momentum and position are related by the Heisenberg Uncertainty Principle, so are energy and time. Deviations from conservation of energy can occur if the deviation stays for no more than a certain minimum amount of time. For example, say that $\Delta E \Delta t \geq 6$ describes the limits on what we can measure (in some appropriate set of units) with ΔE representing our uncertainty in energy and Δt representing our uncertainty in elapsed time. Then the Universe has a small Heisenberg "window" in which it can fiddle, defined by $\Delta E \Delta t < 6$. So, the universe can

4.3. CONSERVATION OF ENERGY

create a particle with 2 units of energy provided the particles lives for less than 3 units of time. Because this product of energy and time is less than 6, then we'd never be able to directly measure its existence due to the Heisenberg Uncertainty Principle, and the Universe could never be "caught" with this extra energy around. If that energy hasn't somehow hidden itself from sight within 3 units of time, it vanishes as though it never existed. In many ways, it never did. If, however, it somehow becomes unobservable before that time limit expires, it can stick around indefinitely.

This is the mechanism by which electron orbits in atoms operate. The energy "transmitted" between a nucleus and an electron exists in the Heisenberg Uncertainty Principle's unobservable region. Both the nucleus and the electron orbiting around it are constantly "broadcasting" this energy, but only the energy that reaches its destination is retained to be observed. Thus, the particles sharing their orbits do not "know" in advance where the other particles are, but the energy exchanged is still 100% efficient. An electron sends out the energy in a field around itself; the energy it loses to the nucleus is exactly balanced by the energy gained by the broadcasts coming from the nucleus. Meanwhile, the energy sent out that doesn't encounter another particle fades into nothingness, and can never be observed, and is therefore not really "lost" by the electron in the end.

So, what exact form does this energy take as it travels? We are used to classifying the world into two categories: particles, which are made out of "stuff," and waves, which are motions of particles. If you have ever dropped an object into a still body of water, you know that waves spread out in all directions.[1] If the broadcast energy was sent in such a wave form, it would also disperse in this fashion, and the energy would then have to find some way to redistribute itself when it encountered its target. Energy spread out over a given area would then have to suddenly rush to the point of contact, increasing the amount of time it spent subject to observation and making it less likely to escape Heisenberg limits. The alternative is much simpler: if the energy was sent in the form of a particle, then 100% of the energy would be carried in a point of zero volume, and would be delivered to the target simultaneously, minimizing the "exposure time" that it spends risking observation and nonexistence. Therefore, we can safely conclude that energy is transmitted in the form of short lived particles, and these particles mediate interactions between other particles, including all forces. We call these particles *virtual particles.*

This quantum mechanical picture of forces, with violations of conservation of energy that can never be observed, can be hard to stomach, particularly if you have already had a few years of physics instruction that drill conservation of energy into you to the point where you cannot imagine an alternative. The rest of this lesson will be spent studying this picture and showing how effectively it really does explain that which we see in the quantum mechanical world.

---

[1] If you have never dropped an object into a still body of water, go find something waterproof, fill up a bathtub and try it. Do it now; this book will be here when you get back.

4.4 The Other Forces

We have already noticed that there must be other forces at work than the two we are familiar with. One force is that which holds the nucleus together; whatever it is, it must be stronger than the electromagnetic force that would push the protons apart. It must also be felt by neutrons, as they are part of the nucleus, but it does not affect electrons, which are never found in a nucleus. Moreover, it must have a limited range, because neighbouring nuclei do not fling themselves into each other overcoming the electromagnetic repulsion they experience from a distance. Similarly, the force which governs radioactive decay (which does involve electrons, and so must be a distinct force) also has a distance limit, since neighbouring nuclei have no influence over the decay of another nucleus. The first question we ask applies to both forces: what could limit the range of a force?

Our forces are mediated by virtual particles who are limited by a maximum value of $\Delta E \Delta t$. We also know that there is a maximum speed in the universe. Let us backtrack our observations and see where they lead:

- The forces in question have a limited range.

- If they have a limited range, the virtual particles mediating the force have a maximum distance they may travel. We have seen no limits on distances particles can travel, so we need to keep digging.

- With a maximum speed and a maximum range, the particle will have a maximum lifetime. We haven't seen any upper limit on the lifetime of a particle.[2] Again, we keep digging.

- If we have an upper limit on Δt, and we have a lower limit on the product $\Delta E \Delta t$, then we are forced to conclude that there is an lower limit on ΔE. In other words, if a particle's lifetime cannot be more than 2 units, and $\Delta E \Delta t$ cannot be less than 6 units, then ΔE cannot be less than 3 units.[3]

There is no reason to expect a minimum speed for virtual particles, or a minimum kinetic energy, so we are led to another seemingly inevitable conclusion: the particles mediating these two forces have mass.[4] By $E = mc^2$, if we use a virtual particle of a particular mass m, then any such virtual particle in nature must have at least mc^2 units of energy. That provides a lower limit on the energy of the created particle.

This conclusion is half right. The force governing radioactive decay, known as the weak nuclear force, (felt by neutrinos) uses mediating virtual particles[5] that

---

[2]Half lives come close to this, by predicting how long an unstable particle can stick around before decay becomes likely, but the random nature of radioactive decay means that there is no limit on an individual particle's lifetime. Thus, the force would be greatly weakened over distance if it used unstable particles, but it would not be completely cut out at a certain point as it actually is.

[3]If $\Delta t = 2$ and $\Delta E = 1$, then $\Delta E \Delta t = 2$ which is below its lower limit.

[4]Conversely, the electromagnetic and gravitational forces must use massless virtual particles, as they do not seem to have any range limit at all.

[5]There are actually three particles that mediate this one, named W^+, W^- and Z_0, where the superscripts refer to their electric charges.

4.5 The Strong Nuclear Force

When researchers first looked for the particle that mediates the strong nuclear force, they expected a particle with mass to account for the limited range of the force. As the strong nuclear force appears to have a more limited range than the weak nuclear force,[6] it stood to reason that the mediating particle would have even greater mass than that which mediates the weak nuclear force.

The particle which mediates the strong nuclear force and glues the nucleus together, called the *gluon*, has no mass at all.

This appears to be a problem. The only explanation we have so far for the limited range of a force is a virtual particle with mass. Well, upon closer inspection, there is a property of virtual particles that is unique to the gluon.

The (as yet unobserved, and therefore technically theoretical) particle which mediates gravity, called the graviton, has no mass. This means that the graviton itself does not feel a gravitational force. The photon, which mediates the electromagnetic force, has no electric charge, and does not feel the electromagnetic force itself. The gluon, however, does carry the charge for the strong nuclear force. This discovery gave researchers a lot of hope: a unique property of a particle means unique behaviour can be expected. They looked into how this property impacts the strength of the force over increasing distances, and found something similarly unique.

As you increase the distance between two particles attracted by the strong force, the gluons mediating the force between them start interacting with each other, creating even more virtual particles along the line separating the two particles. In short, the attractive force between them *increases*.

Suddenly, explaining why two neighbouring nuclei fail to attract each other is not nearly as difficult as explaining why the strong nuclear force fails to pull the Earth into the Sun with an acceleration that puts any amusement park to shame.

The Heisenberg Uncertainty Principle solves this problem, too. As you start to pull particles which feel this force apart, they pull against each other more strongly. It takes more energy to keep them apart. When they are far enough apart, the energy required to hold them apart becomes greater than the energy required to make new particles through $E = mc^2$. That's exactly what happens.

Virtual particles are created between our original two particles. There is enough potential energy between our original pair to account for the virtual particles, so they can become real, permanent, observable particles without violating conservation of energy, and so they do. Because the increased force between the original two particles is confined to the line connecting them, the original pair and the new pair break apart into smaller, enclosed "cells." This not only explains the range limit on

---

[6]The weak nuclear force can, on rare occasion, mediate interactions between electrons and the nuclei they orbit. On the other hand, protons and neutrons that slip into an atom as comparable distances from the nucleus as orbiting electrons do not experience a strong nuclear force.

the strong nuclear force, but it explains why we haven't seen exactly what makes up the internal structure of the protons and neutrons (which we have known was there since chapter two.) The particles protons and neutrons are made of, which are called quarks, are confined to small, complete groups. These groups form protons, neutrons, and less common, more exotic particles we will not discuss in these lessons.

This process works for many forces. The most common way for researchers to create new particles is to impart large amounts of kinetic energy to a small number of particles, and then collide these high energy particles into each other, providing the opportunity for this energy to convert into mass in the form of new particles.

4.6 A New Question

So, our new picture of forces involves the exchange of particles that only get anywhere if they reach their targets. We can almost picture this as though each particle carried around paddles with rubber balls tied to the middle by elastics. Higher energy balls have shorter elastics. If the ball is thrust from the particle without hitting anything, it comes back as though it had never left. If it does reach another particle, it collides, transfers its energy and momentum, and falls limp without returning to the paddle.

However, the particles carrying the balls and paddles, the balls themselves, and the particles that are getting hit by the balls all have zero volume. They are, in essence, impossibly small bulls-eyes to hit. How do they ever reach their targets?

Chapter 5

Let There Be Quantized Electromagnetic Radiative Energy

5.1 Unanswered Questions

1. The basic building blocks of matter, called elementary particles, must all have zero volume. What, then, prevents them from piling in so closely together that the matter they form does not have zero volume?

2. How do we know the Heisenberg Uncertainty Principle applies to particle existence, and not merely measurement?

3. How can particles with zero volume interact through virtual particle exchange at all?

5.2 Some Words About Electric Current

The atomic model used in these lessons so far is akin to a little solar system. Electrons orbit the nucleus like planets at some variety of orbits. If we arrange these atoms near each other and apply a voltage across, it is not hard to imagine the electrons moving like little meteors from solar system to solar system. The electrons closest to the positively charged voltage source leave their solar systems to move toward the attractive charge. This leaves a vacancy in that can be filled by an electron from the next atom, and so forth down the line, until you reach the electrons closest to the negative end of the voltage source trying to escape from that repulsive force.

5.3 Some Words About Waves and Light

"Light" is a word used in the common vernacular to describe illumination we can see. In the science world, it is understood that the visible spectrum is a relatively small portion of the electromagnetic spectrum. Radio waves, microwaves, infrared (heat) waves, ultraviolet waves, X-rays and more are the same essential phenomenon occurring at different frequencies. Waves have a few important aspects that will come into play.

5.3.1 Shape

There are two kinds of waves in nature. Sound travels as a longitudinal waves[1] a series of compressed and uncompressed bits of air are perceived as sound because the changes in pressure cause our eardrums to move, driving the bones behind them to respond. Light, and most of the other waves we'll discuss, is composed of transverse waves.[2] These are the kinds of waves you see on the surface of water: they have high points (crests) and low points (troughs). When two waves overlap, they combine accordingly. If two crests overlap, they form an even larger crest. Two troughs make a deeper trough. A crest and a trough will cancel each other out, reducing the overall strength of the wave. This is called interference; a sufficiently precise setup can produce some extreme examples.

5.3.2 Speed

A given wave will travel at a defined speed. This depends primarily[3] on two things: the type of wave, and the material it is waving through. In other words, all sound waves in air travel at the same speed, regardless of pitch.[4] If those same sound waves hit water, they'll travel through the water at different speeds than in air, but all sound waves in the water will travel at the same speed.[5] The same is true of light in all its forms: radio waves, visible light and X-rays all travel at the same speed through the same medium.[6]

---

[1] Etymology of "longitude": along the distance or direction of motion
[2] Etymology of "transverse": lying across: waves travel across the direction of motion.
[3] Frequency and other variables have a slight effect on it, but very slight.
[4] Those of us who can only afford concert tickets at large distances from the stage really appreciate this. Concerts would be much less enjoyable if vocals, bass, lead guitar and drums were out of sync by the time the sound reached us.
[5] The speed does depend on the temperature of the medium, meaning the speed of sound in warm air differs from the speed of sound in cold air, but that's still a property of the medium, not the wave.
[6] Note that Einstein's relativity specifically sets the universal speed limit as the speed of light in a vacuum, where light travels most quickly. When light is travelling through another material, it slows down. As a result, it is possible to travel faster than light, if that light is travelling through a medium. For example, high energy electrons can outrun light in water, and produce something called Čerenkov radiation in the process. This is effectively light's version of a sonic boom.

5.3.3 Wavelength

The wavelength is the distance between identical parts of a wave. With transverse waves, it is most easily measured as the distance between two crests or two troughs. (The distances will match.) With longitudinal waves, it is the distance between two consecutive compressed or uncompressed portions of the wave. This property depends on both the medium and the wave itself: all light that is a particular shade of blue will have the same wavelength in air, but when that light moves into water, all light that shade of blue will take on a different, but equally specific wavelength.

5.3.4 Frequency

Frequency is another word that has common and scientific meanings. In this case, the meanings are more similar than they are with the word light.

Frequency commonly means how often something happens. That's what it means in science, too: we just have specific definitions of the "something." The frequency of a wave is how often a complete wavelength of the wave passes a specific point.

The frequency is the characteristic that most scientists use to describe waves, simply because it is the characteristic that depends solely on the wave, and not on the medium. Different colours of light will have different frequencies, but those frequencies are the same whether that light is travelling through air, water, glass, vacuum or transparent aluminum.

Frequency can also be used to tie wavelength and speed together. Imagine we have a wave with a frequency of three wavelengths per second[7] and a wavelength of two meters. Combining these two pieces of information tells us that a single point on the wave will travel a total of six meters in one second. Thus, the speed of a wave can be found by multiplying the frequency and wavelength together.

5.3.5 Intensity

The intensity of a wave is the amount of that wave coming in. This depends on the energy output of the source: the Sun has a much greater intensity of light than a light bulb that glows the same colour. You can double the energy output by doubling the intensity without changing the frequency, speed, or wavelength of light. Intensity of light is perceived as brightness, while intensity of sound is perceived as volume. Thus armed with the fundamentals of wave theory, we are ready to proceed.

5.4 The "Neener Neener" Spot

To this point, we have not addressed the nature of light in these lessons. Historically, it was believed that the wave nature of light was well understood by physicists. One of the most compelling arguments that light was a wave, not a particle, came in

---

[7]The most common unit of frequency is the Hertz, named after a major contributor to classical wave theory, where 1 Hz means one wavelength per second.

the form of the Poisson spot, which may be the only phenomenon in the realm of science that was named after the individual who did not find it.

Simeon Poisson was a physicist in the early 1800s who was known for having more talent with theory than experiment. In fact, his experiments had an uncanny knack for producing exactly the theoretical result Poisson expected. While physicists were hotly debating whether light was a particle or a wave, Poisson designed an experiment that would seem to settle the debate.

Imagine a calm river. Some object, such as a rock or fallen tree, is protruding from the river's surface. This will create waves and ripples as it disrupts the natural flow of the river. These waves and ripples will wrap around behind the rock, often creating an area of even higher waves than the surrounding area. The same is true for light waves.

If one were to mount a circular object near a light source, then one could set up a similar situation. If light is a wave, and if it falls on this circular object, then the wave could wrap around the obstruction and "add up" on the other side to produce a bright spot in the middle of the shadow. (The observation screen would have to be placed with high precision, but it's possible.) This bright spot would be required if light were a wave, and would be completely inexplicable if light were a particle.

Poisson was a supporter of the particle theory of light, and when he performed the experiment, he found no such bright spot. He promptly declared that light was a particle. However, Dominique Arago knew of Poisson's history of finding exactly the results he expected, so he repeated the experiment and found that the spot was, indeed, right where wave theory predicted it would be. He promptly named the bright spot the "Poisson Spot," simply to rub it in.

Nonetheless, the discovery of the Poisson spot proved beyond a shadow of a doubt that light behaves like a wave.

5.5 The Blackbody Radiation Problem

By the early 1900s, there were a couple of problems with phenomena involving light. One was the so-called blackbody radiation problem.

A blackbody is an object that is entirely black in colour. A quick experiment with most stoves will reveal that heating a black object to a high enough temperature causes it to glow. The glow is initially red in colour, but as the temperature increases it becomes more yellow, and eventually white. Experiments had verified this pattern is independent of the material involved. Any object heated to a given temperature will glow a particular colour. In other words, it produces light.

Herein lies the problem: in the classical model, light waves were produced by a vibration of the material itself. In that model, a piece of the blackbody would vibrate with heat energy and emit a wave. The wavelength of light would be determined by the heat energy causing the vibration in the original material.

The problem is that shorter wavelengths would be more common than longer wavelengths. Classical theory predicts that most light would be produced in the X-ray and ultraviolet regions, and that the first visible colour as the temperature increased would be violet, not red. Any time theory doesn't match experiment,

5.6. THE PHOTOELECTRIC EFFECT

it's time to review the theory. The blackbody radiation problem was not the only problem with classical theory.

5.6 The Photoelectric Effect

A second significant problem with classical theory came from the photoelectric effect. This is the phenomenon that drives solar powered calculators and similar technology.

There are some materials in nature that can act as sources of electric current when exposed to light. If you were to put such a material in an electric circuit with no voltage source and shine a light on it, current would flow. At this rudimentary level, the ideas we've established so far are more than enough to explain the phenomenon: light energy is imparted to electrons in orbit, and this energy is enough to cause the electrons to break orbit and travel through the circuit.[8] When we look more closely, the existing theory is proven inadequate. Existing theory would indicate that, once the photoelectric effect is established as a real phenomenon, the relationship between light and current would be simple. Doubling the intensity of light would double the current. Doubling the frequency of light would double the stopping voltage. Neither of these effects were seen.

Problem #1: The light frequency impacts the energy of the electrons in the current. Each electron gains energy from exposure to light, and starts moving in a current. If we add a battery to the circuit which opposes the current, we find that there is a particular voltage level that brings the current to a complete stop. The inexplicable part is that increasing the frequency by 20% also means increasing the voltage, but not by 20%. For example, if the frequency changes from 1000 Hz to 1200 Hz, the voltage may change from 10 V to 20 V. Even more strange is that the absolute difference is maintained; in the above example, if you increase the frequency by another 200 Hz to reach 1400 Hz, then the voltage needed to stop the current (called "stopping voltage") increases by the same 10 V difference we had before, bringing us to 30 V. A frequency of 2000 Hz would drive this up to 60 V; a 100% gain in frequency caused a 500% gain in stopping voltage.[9]

Problem #2: In order to produce the photoelectric effect, a certain minimum frequency of light is required. No intensity can overcome this: just because photoelectric current flows when the photoelectric material is exposed to a given intensity of light with a frequency of 2000 Hz, there is no guarantee that current will flow with double the intensity of 1000 Hz light. In fact, there is no guarantee that current will flow at all. Different materials have different minimum frequencies.

Problem #3: Increasing the intensity of light increases the current, but not the stopping voltage. If twice as much light falls on the material, twice as many

---

[8] Electrons were originally discovered as a result of the search for the moving part of electric currents. It was already well known that current was nothing more than electrons in motion.

[9] These are entirely fictitious numbers used for the purposes of explanation only. An increase of 200 Hz leads to an increase of 8.27×10^{-13} V for electrons in all photoelectric materials. This independence of the relationship between materials could be a fourth problem in its own right, but it doesn't seem to be one that greatly concerned the original experimenters, possibly because the solution was found before this was seriously pondered.

electrons flow, but each electron carries no more energy it did under the original intensity at the same frequency.

These were extremely perplexing problems, at least until they were studied by a guy named Albert Einstein.

5.7 Quantization of Light

Albert Einstein did not win his Nobel prize for the theory of relativity. Instead, he won it for finding a single explanation for the three problems with the photoelectric effect, and pointing out that the same explanation would also take care of the problems with blackbody radiation.[10] The solution is simple: treat light as a particle.

Ignore the Poisson spot for a moment.[11] Imagine that light is a particle, and that each particle contains a particular quantity of energy. Planck's math had already indicated that this energy would be directly related to the frequency. He refused to believe the quantized light interpretation because of the Poisson spot, but he published anyway because the math worked so well. In his model, energy E and frequency[12] ν of light are related by $E = h\nu$, where h is a constant that exists mainly as a conversion factor between units. A particle of light of a given frequency (or colour) will have a given amount of energy.

Now imagine these particles of light energy falling on a material loaded with orbiting electrons. The electrons are able to absorb this light as energy, but (it seems) they can only absorb one particle at a time. If each particle doesn't have enough energy to get the electron out of orbit around the nucleus, then the electron doesn't go anywhere, and no current is produced. Instead, the energy is either retained as heat, or released as reflected light or blackbody radiation. No intensity of low frequency light will cause an electron to escape atomic orbit and become electric current: this takes care of problem #2.

Once we've hit the threshold frequency, or the frequency of light that is high enough to cause electric current to flow, we can explain the other two problems. If we have photoelectric current and we double the intensity of light, we double the current but don't change the stopping voltage. This also makes sense: with twice as many light particles falling, we can free twice as many electrons to flow, but each individual electron's energy is still limited by the energy of the individual light particles, so the stopping voltage doesn't change. Problem #3 is no longer a problem.

Finally, we have our original problem to deal with. The relationship between stopping voltage and frequency is not strictly linear, though there is a definite and

---

[10]The mathematics that back it up had already been done by Max Planck to deal with a separate issue.

[11]Einstein did. He had to. If you hold on to the wave model of light the Poisson spot had definitively proven, he could not have explained reality.

[12]In most high school courses, frequency is denoted f. The professionals use the Greek letter "nu," ν, and they use it consistently. This tends to confuse students in their first year of post-secondary with what appears to be a sudden and inexplicable change in notation: the post-secondary institution didn't change the notation, the public school did.

5.7. QUANTIZATION OF LIGHT

distinct pattern. Planck's formula comes to the rescue on this one, in a fashion. The solution to this problem will lead to an entirely different pair of problems.

It was well known before the photoelectric effect was fully explored that a charged particle can gain kinetic energy in a circuit if it moves through a voltage difference. If you apply a voltage V to a particle with charge q, then the energy gained (or lost, if the voltage tries to push opposite to the direction of motion) is equal to $q \times V$. Since this is an observable phenomenon, conservation of energy applies.

The total energy used came in through the original particle of light. Some of that energy is used to get the electron out of its orbit, and the rest can be used to make the electron move. This explains things nicely; there is a linear relation with uniform increases because this energy is partitioned into two parts. One part is the energy required to get the electron out of orbit, and the other part is the energy the electron uses to move through the circuit. If we take the energy of the electron moving through the circuit[13] and add the energy needed to get the electron out of orbit, called the work function, we get exactly the energy the incoming light particle had by Planck's math.

The blackbody radiation problem is also neatly explained by this solution: to produce the higher frequencies of light that we seem to be missing, we would need to impart a relatively large amount of energy into them. This caps the amount of light particles that can be produced at higher frequencies. Rigorous math proves that this modification is exactly what is needed to produce blackbody radiation in the colours we see at the temperatures at which we see them.

When two distinct problems can be solved by a single, relatively simple idea, that's usually considered definitive proof that the idea is correct. In this case, Einstein proved beyond a shadow of a doubt that light behaves like a particle. We call these particles photons. That term has appeared before; photons are the particles that mediate the electromagnetic force.

5.7.1 Reconciliation Problem

Now we have seemingly contradictory results. The existence of the Poisson Spot proves, beyond a shadow of a doubt, that light behaves like a wave. Now, Einstein has proven, beyond a shadow of a doubt, that light behaves like a particle. This contradiction will need to be reconciled.

5.7.2 Work Function Implications

When we graph most physical phenomena, we usually get smooth curves in the graph. Even rapid changes produce curved graphs, although the curve may be small and tight.

When you graph the frequency of light incident on a photoelectric material against the current that flows out of it, you do not get a curve. When you hit the work function, you get a sudden, sharp change in the current flowing, as though

---

[13]The energy of the electron moving through the circuit is equal to the charge on the electron multiplied by the voltage required to stop it. That's the amount of energy we had to take away from the electron to make it stop, so that must be the amount of energy it had to begin with.

someone hit a switch. This is a different kind of phenomenon than anything we have encountered before.

Beyond that, this also means the work function is extremely well defined. This is not a random number, so electrons are not arranged in the materials as randomly as planets are arranged in a solar system. The atomic model we have been using so far needs to be adjusted, as does our picture of electric current.

Chapter 6

Quanta, Quanta Everywhere

6.1 Unanswered Questions

1. The basic building blocks of matter, called elementary particles, must all have zero volume. What, then, prevents them from piling in so closely together that the matter they form does not have zero volume?

2. How can we reconcile the Poisson spot with the photoelectric effect and blackbody radiation?

3. Why does the work function kick in so suddenly in the photoelectric effect?

4. If the work function is so well defined, how are electrons arranged within a material?

5. How does electric current really work?

6. How do we know the Heisenberg Uncertainty Principle applies to particle existence, and not merely measurement?

7. How can particles with zero volume interact through virtual particle exchange at all?

In this lesson, we focus on questions #2 and #6, and the implications of their solutions.

From question #2, we have contradictory results to reconcile. There is no question that light behaves as both a particle and a wave. This was a revelation. The first step in the investigation of this problem was to discover if light was the only object in nature to suffer from this problem.

6.2 Diffraction

To investigate whether or not other particles behave like waves, we must first establish some properties and behaviours that are unique to waves.

The first such behaviour is diffraction. As mentioned in the previous lesson, waves can interfere with each other. In the case of light, that interference produces bright and dark spots in a regular pattern, depending only upon the geometry of the situation and the wavelength of the light.[1] The setup requires a point source of light, a screen to observe the light, and some sort of physical block or impediment that ensures light falling on the screen has as least two discrete paths to follow to arrive at the final destination. This can be done by putting an opaque screen with two slits between the light source and the screen. Alternatively, it can also be done by setting up two mirrors with less than 100% reflection, allowing light passing through to reflect off of either mirror.[2]

Imagine that a wave of light passes through the longer path, and arrives at the screen when its wave is at a crest, or the highest point. Now imagine a second light wave travels the shorter path, and arrives at the screen at exactly the same time as the first wave of light. If that second wave is also at a crest, then the two will combine to make a bright spot. If that second wave arrives at a trough, on the other hand, the two will cancel each other out, producing a dark spot. This not only shows the wave behaviour of light, but it means that an analysis of the complete pattern resulting can determine the wavelength of the light involved. (Bright spots appear when the difference in distance travelled by the two waves along two paths is a whole number of wavelengths. Dark spots appear when the difference is half a wavelength more than a whole number of wavelengths.)

In 1927, Lester Germer and Clinton Davisson set up an experiment similar to the mirror setup described above. Instead of mirrors, they used nickel, which reflects electrons. Instead of a light source, they used a source of electrons. Instead of a screen, they used a device that counts electrons.

They found places where they counted a lot of electrons, and they found places where they counted very few electrons. In short, electrons combine with each other like waves. Instead of being brighter or darker, there are more and less of them.[3]

This is staggering. It has also been reproduced for protons, neutrons, and any other particle we have examined this way.

The wave nature of the objects we are used to calling particles is not only present, but it seems to impact whether or not the particles even exist. The number of electrons launched at the nickel, the total charge on them, and their total kinetic energy are all observable quantities, so they must conform to conservation laws. The electrons have to go somewhere, and they do. Somehow, some way, these little zero volume billiard balls we have been picturing move through space as though they have already chosen their destinations based on where other particles are in the same situation.

This gets into an information exchange problem. Again, the particles seem to

---

[1] The light used for this must be of a single colour, or a single wavelength. White light and so forth can get too muddled to see the effects clearly in many cases.

[2] These are certainly not the only arrangements, but they are among the most popular and easiest to set up.

[3] By this stage, it probably comes as no surprise to learn that "eureka!" is almost never spoken aloud in physics research institutions. However, it is very common to hear variations of "what the *expletive* is *that*?"

6.3. PARTICLE OR WAVE? 35

demonstrate precognition, arriving where they need to arrive to coincide with other particles. We have already developed a means of information exchange between particles, but it ties directly to forces. The forces involved with these particles are not consistent. We know that electrons repel other electrons, and yet they are clumping together at the same destination. We know that the only long distance force neutrons experience is gravity, and that force is not nearly strong enough to account for this behaviour. No, the particles cannot be exchanging information. There must be another answer.

6.3 Particle or Wave?

The first attempt to reconcile these ideas was something called "wave-particle duality." Scientists realized that there was no way to treat subatomic objects as either particles or waves. These entities exhibit properties of both. The solution was to propose that they are both. Every basic building block in the Universe is both a particle and a wave, exhibiting all properties of both when placed in situations where those properties matter.

As hard as this was to conceptualize, it did provide the logical consistency required for scientists to accept things as they saw them. This explains much of what we see, but not all. It explains why light behaves like a particle, and also why electrons have the wavelength necessary to demonstrate diffraction. It does not, however, effectively deal with the problem of a precognitive electron waving to exactly where it needs to be: the number of electrons we count at the destination still varies as though the electrons already know where to go.

6.3.1 None of the Above

The proper solution to the problem is even more difficult to grasp. Electrons, photons, protons, neutrons and all of their ilk are neither particles nor waves, nor are they any combination of the two. They are something else, something entirely different, which has no macroscopic analog that we can relate them to from our day to day experience. We need to develop an entirely new mental picture to understand what these entities truly are. To that end, as promised in lesson three, it is now time for Heisenberg to make his triumphant return.

6.4 Heisenberg and Existence

If the electrons, protons, and other particles that exhibit wave behaviour determine their destinations before arriving at them, then we find ourselves stuck with precognition and forceless information exchange. However, they certainly react to other particles travelling similar journeys; we know this because they behave like waves.

The solution to this seeming discrepancy is as bizarre as the problem itself: the determination of where particles "arrive" at the detector isn't made until they get there. Along the way, these particles are not the zero volume particles we have deduced them to be. Rather, they are regions in which it is probable to find said zero volume particle when you look for it.

These regions of probability are the ones that demonstrate the wave behaviour. When we put the electron counter in place, and force an interaction that depends upon the precise current location of the particle, then the region "collapses" to a single point which represents our zero volume particle. There is no time required for this collapse, either, so the constraints on the wave image of information transmission from lesson four do not apply. Instead, the Universe behaves as though the particle had always been exactly where the region collapsed in the first place.

When calculations were done for this model,[4] they demonstrated that the indeterminate regions follow identical limitations to the Heisenberg Uncertainty Principle. This "fuzziness" we need in the existence of our particles exactly matches the fuzziness in our ability to measure and detect them. This is the proof we were waiting for to justify the statement that the Heisenberg Uncertainty Principle applies to existence as well as measurement.

This is how we answer the last unanswered question on this chapter's list: zero volume particles can interact with each other because the virtual particles they exchange do not need to reach a zero volume particle, but instead need to reach a region of space with more than zero volume. When two such fields overlap, then there is a probability of interaction, and this is the interaction that is perceived as a force.

6.5 Inverse Square Laws

Now that we have established that zero volume particles still have physical extents when not being observed, we can establish another important characteristic of the electromagnetic, gravitational, and weak nuclear forces. This section is more mathematically involved than most, and is meant to explain a phenomenon to people familiar with the particulars of a force. The most important point here is that the force between two charged objects changes more rapidly if we change the distance between them than if we change the charges on them. Readers who are not mathematically inclined can ignore the rest of this section without impacting the readability of the remaining chapters.

The electromagnetic and gravitational forces both have the same basic algebraic structure:

$$F = \frac{cp_1p_2}{r^2}$$

In this notation, c is a constant that is representative of the strength of the force in general,[5] p_1 and p_2 are properties representing the specific strength of the force

---

[4]The calculations were done with the intent of demonstrating the utter ridiculousness of the model so that it could be thrown out and replaced with something sensible. This was common. It was also common to perform experiments to prove the utter ridiculousness of these theories, only to find the rules that govern the Universe are at least as ridiculous as mankind can imagine. More on this next chapter.

[5]$c = G = 6.67 \times 10^{-11}$ N·m^2/kg^2 for gravity, $c = k = \frac{1}{4\pi\epsilon_0} = 8.99 \times 10^9$ N·m^2/C^2 for electromagnetism

6.5. INVERSE SQUARE LAWS

for the two objects experiencing the force,[6] and r is the distance between the two objects.

The question is this: why do the forces always take this particular form? The properties in the numerator match intuition: if you double the property that generates the force, perhaps by doubling the mass of one object, then you double the strength of the force. However, the less intuitive part is the exponent associated with the distance between them. If you have two masses a certain distance apart, double the mass of one and then double the distance between them, our intuition would often tell us to expect the force to remain the same. Instead, the force between them is cut by half. We refer to this as an inverse square law: the force depends on the inverse square of the distance.[7]

6.5.1 Classical View

In the classical view, the exponent of r turns out to be rather easy to explain, provided you can imagine a sphere. In this view, each object produced a field. For example, let us imagine the two objects are the Earth and the Moon, and that the force between them is gravity. Earth would produce a gravitational field, which would be perfectly uniform in all directions if Earth were a perfect sphere.[8] The field would then be distributed through space evenly in all directions. As you move farther from the Earth, the field would be spread out more thinly.

Now let us imagine the Moon being placed in this field a distance r away from the Earth. The Moon is roughly spherical: when viewed from the Earth, it looks very much like a circle. The strength of the force it will experience is then going to depend on the proportion of Earth's gravitational field that it experiences. If the circular face of the Moon has an area A_{Moon}, then the strength of the force it experiences will depend on the percentage of the Earth's gravitational field that it inhabits. This percentage is determined by the ratio of the Moon's area to the total area in space that the Earth's field is spread over (A_{field}), which (mathematically speaking) is this ratio:

$$\frac{A_{Moon}}{A_{field}} = \frac{A_{Moon}}{4\pi r^2}$$

where we have used the fact that the area of a sphere of radius r, which the Earth's field is spread over, is given by $4\pi r^2$.

This is where the r^2 dependence comes from in the classical view. The geometry that deals with the 4π and other components gets absorbed mathematically in the properties (m or q) of the charges and the constant (c) in the force equation.

---

[6] $p = m$ for mass or inertia with the gravitational force, $p = q$ for electric charge for the electromagnetic force

[7] The "inverse" part means r appears in the denominator, or bottom, of the fraction instead of the top, while the "square" part refers to the exponent 2.

[8] Earth actually bulges at the equator due to its rotation. This also means that you will lose a bit of weight if you move closer to the equator, without changing your mass, simply because of this same r^2 dependence in the denominator of our force equation. The average distance between yourself and the centre of the Earth would increase closer to the equator. You will lose more weight if you walk or run to the equator in the process.

This view is based on the classical notion that we can model the field of one object in isolation, and then add a smaller object to the system without significantly impacting the field produced. In our quantum mechanical view, we understand that forces only exist when you have two entities involved, which can both send *and receive* virtual particles. Thus, you cannot create a field with an isolated particle when no interactions are possible, and the virtual particles that mediate the exchange do not distribute themselves in a perfect sphere, but rather appear in random directions and only matter when they interact with other particles. We need a new mental picture to explain the inverse square forces.

6.5.2 The Quantum View

As a first attempt at building a view of inverse square laws with our quantum view, we could attempt to say that the extent of our region of probability, or field of influence, acts like the visible area of the Moon acted in the classical view. This seems acceptable at first, but fails when we realize the virtual particles are not spherically distributed unless the particles they can react to are spherically distributed.

To explain things at the quantum level, we need another picture. The forces only exist when the objects exist in pairs, so we cannot model one field and then insert a second object into it.

Recall that the virtual particles exist within the limits of Heisenberg Uncertainty, which (as we can now confirm) limits existence as well as measurement. Let us also recall that the energy transferred by these virtual particles only "counts" if the particles reach a destination, at which time they are observable and subject to conservation of energy.

The particles of energy E that reach their destination within time t are then constrained by $Et < k$, where k is some constant about the situation we do not yet know. (It will vary from any given situation to any other, and cannot be exactly determined for a general discussion.) Keep in mind, also, that the Universe enforces a speed limit of the speed of light in vacuum. That means there is a limit to how far a particle can travel in time t. If we exchange high energy particles, then they can only be exchanged over short distances, because of this speed limit. If we exchange low energy particles, then they can be exchanged over larger distances, because a longer transit time is allowed by Heisenberg.

This is the dependence we need. The virtual particles contributed to the force by one real body will fade smoothly as the distance is increased. If the energy of the virtual particles produced is directly related to the charge property (p) of the real particle, then each real particle contributes a factor of $\frac{p}{r}$ to the overall force. In this case, the force equation for our quantum mechanical view becomes

$$F = c\left(\frac{p_1}{r}\right)\left(\frac{p_2}{r}\right) = \left(\frac{cp_1p_2}{r^2}\right)$$

because the distance r between the real particles is identical from each particle's perspective.

This is exactly the same algebra that was produced by the classical view that is consistent with experimentation, which is what we needed. We can then explain

6.5. INVERSE SQUARE LAWS

the forces experienced by large objects, like the Earth and the Moon, by adding up the influences of every individual particle involved. Thankfully, the calculation involved in the sum is exactly the same as doing a single calculation looking at the total charges between the two summed over all particles involved in both bodies, using the distance between their centres as the total distance between them, which is exactly the classical formula.[9]

---

[9]The alternative would be to perform over a googol's worth of calculations between each particle in the Earth and each particle in the Moon and add up the results.

Chapter 7

Down the Rabbit Hole

7.1 Unanswered Questions

1. The basic building blocks of matter, called elementary particles, must all have zero volume. What, then, prevents them from piling in so closely together that the matter they form does not have zero volume?

2. Why does the work function kick in so suddenly in the photoelectric effect?

3. If the work function is so well defined, how are electrons arranged within a material?

4. How does electric current really work?

5. How can a particle possibly be some sort of probability field?

7.2 Welcome To Wonderland

Last lesson, we developed a truly bizarre view of subatomic particles. In this view, they are regions of space corresponding to the probability that a particle exists at a given point within that region, and only becomes a single point particle when it interacts with its environment.

The normal human reaction the first time this idea arises is abject rejection. The concept is so far removed from typical human intuition that it is almost irreconcilable for many of us. There are two famous quotes which basically sum up the human reaction to what has come, and to what it still coming:

> Anyone who is not shocked by quantum theory has not understood a single word. (*Niels Bohr, 1885-1962*)

> Baby, you ain't seen nothin' yet. (*Bachman-Turner Overdrive, 1973-present*)

7.3 The Rabbit Hole Without the Hole

One of the earliest implications of this idea is known as tunnelling, and it is one of the fundamental differences between the behaviour we see in the macroscopic world and the microscopic world.

For now, let us look at the macroscopic model. Imagine you are faced with a wall. Your goal is to get the ball in your hand to the other side of the wall. The wall is high enough that you cannot throw the ball over. The wall spreads out to the sides farther than the eye can see, so you cannot go around.

The theories we are used to say that you cannot get the ball to the other side of the wall. This new "region of probability" picture of quantum mechanics says that, if you throw the ball at the wall, there is a small chance that the ball will tunnel directly through the wall, appearing on the other side without harming the wall in any way.

Upon realizing this implication, physicists went charging to their labs, hoping to disprove this insane mental picture of the quantum mechanical world so that it might be replaced with something more sensible.

They failed spectacularly. When you create this situation in the quantum mechanical world, using an electron, photon, and so forth as your ball, you find that some of the balls get through. Moreover, the experimentally determined probability that a ball tunnels through the wall is exactly what this insane theory predicts.

Quantum mechanical particles can and do tunnel through seemingly insurmountable obstacles.

A baseball cannot tunnel through a wall very often.[1] It cannot get over because of conservation of energy; if it does not have enough energy to overcome gravity long enough to reach the height of the wall, it simply cannot get through.[2] As we have already seen, quantum mechanical particles can violate energy conservation if they do so in a way that makes the process undetectable until it is too late.

This is exactly how quantum mechanical particles tunnel. When they hit the quantum mechanical version of a wall, they respond to the change in energy levels. The greater the energy required to overcome the obstacle, the less likely they are to succeed. As long as they have enough observable energy to exist on the other side of the wall, and as long as we are incapable of observing what happens within the wall while the particles tunnel, they can tunnel through.

There is an even more unexpected flip-side to this bizarre phenomenon.

7.4 Through the Looking Glass. Or Not.

Quantum mechanical particles can tunnel through a wall because of the way they behave at a change in energy. Walls are not the only objects that change required

---

[1] We do not consider walls with windows at this point.

[2] Technically, a baseball is a collection of quantum mechanical particles, and there is definitely a very slim chance of success. The probability of every particle in a baseball tunnelling through the wall and coming out the other side looking like a baseball is less than the probability of playing the lottery and winning the jackpot every single week of your adult life.

energy.

Let us return to our macroscopic analogy. You are standing on a sidewalk, and you want to throw your ball across the street. In the middle of the street is a sewer, and the manhole cover has been removed. You do not anticipate any trouble throwing your ball across the street, over the empty manhole.

In quantum mechanics theory, the particle reacts to the change in energy here as well.[3] The reaction is even more strange: either edge of the manhole, which both represent changes in energy, can cause the quantum mechanical ball to reflect back upon its original path. In other words, if you throw a quantum mechanical baseball across the street, it reacts to the lip of the manhole as though it hit a wall.

Again, experimentalists thought they had an impossible situation that could be used to derail this insane train of thought. They rushed to their labs and started launching quantum mechanical balls over quantum mechanical holes.

The balls bounced back, from both edges of the holes. They did so just as often as the theory predicted. It was beginning to look like these crazy ideas were here to stay.

7.5 A Window In The Quantum World

There was one other absurdity in this quantum theory that gave classical thinkers hope. These three phenomena came as a set: if you have regions of probability instead of particles, then these regions behave oddly. The region can tunnel through obstacles, reflect at holes, and be in two places at once. If any one of these phenomena could be disproved, then it meant going back to the drawing board and looking for a replacement theory. Any such theory would now need to include tunnelling and reflecting, but it was hoped by many that a less insane replacement theory would be found.[4]

Let us go back to our wall. We are still going to try throwing the ball to the other side of the wall, but this time, we are going to install two windows in the wall first. Now it is easy to get the ball to the other side: aim for one window or the other. We will start with open windows to prevent property damage.

With a macroscopic ball, you can observe where it lands, and deduce with certainty which window it came through, particularly if those windows are relatively far apart. Microscopic balls are regions of probability, who spread out widely enough that it is difficult to determine exactly which window the ball went through. To complicate things further, microscopic balls have wave properties within this region, and they interfere with each other as when we discussed diffraction, so if you throw

---

[3] Why does an open manhole represent a change in energy? It comes down to something called potential energy, or the energy that can be used to change a system. If you drop a ball from shoulder height to the ground, it picks up a certain speed before it lands. If you drop a ball from shoulder height into an open manhole, it will enter the manhole at the same speed it would have hitting the ground, but then continue to accelerate until it finally reaches the bottom. Thus, when an object is above a hole, it has greater potential energy, as it has the potential to reach a greater speed when in free fall.

[4] We now realize that, in most cases, when a theory fails to describe the quantum world it will likely be replaced by a theory that is more insane, not less.

a lot of balls at the windows at once then this diffraction pattern will emerge on the other side. Physicists looked for this, and they found it. To study the phenomenon further, they decided to slow down the rate they threw the balls at the window. If they slowed things down enough, they could guarantee that the balls went through the windows one at a time, and the pattern of interference would be what they expect from a single wave, rather than a combination of waves.

The combination interference pattern stayed. Even when coming in one at a time, it seemed that there were electrons interfering with each other.

The scientists took the next logical step: they closed the windows. If the balls went through a closed window, the window would break, and we would know that a ball had gone through. If the quantum mechanical windows were rigged to repair themselves, then the window breakages could be used to count incoming particles, and they could determine where the extra electrons were leaking into their experiment.

There were no extra electrons. Moreover, as soon as they closed the windows, the pattern of interference instantly became the pattern expected from a single wave, rather than a combination of waves.

If you do not have a system in place to determine which window the particle goes through, it goes through both, and interferes with itself.

This phenomenon solidified the "region of probability" interpretation of quantum mechanics. There was no other reasonable way to explain observations. It was time to stop fighting the idea, and instead work with the idea to see where it would lead.

7.6 Electron Orbits

We have already established that the manner in which electrons orbit cannot be exactly the same as planets around a star. The specific and sudden work function from the photoelectric effect was a strong indicator that the energy of current carrying-electrons is in no way random, and that these electrons (at least) are contained in the material with some sort of predetermined structure.

As mentioned earlier, incandescent light bulbs work through blackbody radiation. Fluorescent light bulbs were not mentioned; they do not work through blackbody radiation.

Early on, scientists noticed that some gases (such as hydrogen) glow when a current runs through them. This could not be explained with classical physics. Even more confusing, when the glowing light was passed through a prism, it did not produce a continuous rainbow as most light does, but instead produced a series of specific colours. Despite numerous attempts, classical physics failed to describe this phenomenon in any useful way.

Erwin Schrödinger[5] developed a working solution, and he needed this quantum mechanical picture to do it.

In our ball analogy, the potential that determines energy levels is based on gravity. At the quantum mechanical level, the effects of gravity are insignificant.

---

[5]Erwin Schrödinger: excellent physicist, lousy veterinarian.

7.6. ELECTRON ORBITS

Schrödinger instead developed a model of the atom in which the nucleus is the source of electromagnetic potential, and an orbiting electron is a wave-like region of probability that falls into the potential hole produced by the nucleus.

Trapped in this hole, the electron starts reflecting off of the edges, and interfering with itself inside the span. This interference tends to reduce or eliminate the probability of finding the electron at certain points in this well. In fact, only certain orbits are allowed.

When a wave reflects off a hard surface, part of the wave reverses. The wavelength and frequency do not change, but the local shape changes: crests become troughs, and vice versa. Schrödinger found that this can produce a model of electron orbits that explains much of what we have seen. Imagine that exactly half a wavelength of an electron fits in the hole. In that case, the consecutive crest and trough of a single wavelength will reflect back on each other and overlap. However, because the reflection turns troughs into crests and vice versa, the two halves of a single wavelength match (both crests, or both troughs) when they reflect upon each other and overlap in this fashion. They reinforce each other in these orbits, but cancel each other out in others. Similar phenomena happen when the electrons can fit $1\frac{1}{2}$ wavelengths in the well, or $2\frac{1}{2}$, or $3\frac{1}{2}$, or any "whole number plus half" wavelength. Due to difficulty in the math,[6] Schrödinger could only solve the problem for a single electron atom, which means hydrogen, or helium missing one electron, or lithium missing two electrons, or ununhexium missing 115 electrons, and so forth.

The wavelength of an electron depends on its energy. That means that there are preferred energies for electrons to have while they are in these orbits. When calculating the differences between the energies of these different allowed orbits, Schrödinger realized that they are exactly the right differences in energies to match the energy of light emitted when a current runs through hydrogen.

The idea that electrons exist as regions of probability with wave-like properties not only predicted some bizarre behaviours as described earlier, but it perfectly explained the spectra of light emitted from a gas subjected to electric current. The current imparts orbiting electrons with enough energy to jump up to higher orbits. Electrons then "fall" to lower orbits, emitting energy in the form of a photon of light. That photon then contributes to the glow of the light.

This is a huge building block for many of our unanswered questions. Electrons are not only allowed to exist in orbits that line up with the self-interference described above, but these are the only orbits they are allowed to use. Instead of orbiting a nucleus like little planets, they extend through a region like a wave, and these regions come in some very bizarre shapes that will be discussed in our next chapter.

---

[6]This is not to say Schrödinger was a poor mathematician. The math is what we call a "transcendental equation." This means that, if the variable you are interested is x, then there is no possible way to manipulate the equation to turn it into $x =$ stuff that doesn't involve x. It's not just that he couldn't do it, but that nobody could do it. Instead, we solve equations of this type using highly sophisticated "guess and test" methods, where each successive guess leads to a better guess until our final guess becomes close enough for the purpose at hand.

Chapter 8

One and One and One is Three

8.1 Unanswered Questions

1. The basic building blocks of matter, called elementary particles, must all have zero volume. What, then, prevents them from piling in so closely together that the matter they form does not have zero volume?

2. Why does the work function kick in so suddenly in the photoelectric effect?

3. If the work function is so well defined, how are electrons arranged within a material?

4. How does electric current really work?

5. What kinds of orbits can electrons have?

8.2 Volume of Matter

As we learn more and more about the subatomic world, we solve some problems with our models, but introduce others just as quickly. One question which has been lingering since our first lesson can finally be addressed: if matter is made out of pieces with zero volume, then why do any objects have volume? In most public school models, atoms fit together like little billiard balls, so the size of these billiard balls determines how closely the atoms can be packed together. We cannot depend on this logic any more. If atoms are made of objects without volume, why do they align themselves into arrangements which do have volume? In short, how do isolated atoms come together to form the world we live in?

8.3 Molecule Formation

The simplest combination of two or more atoms is called a molecule. This arrangement has a definite microscopic start and end point, and is the easiest to imagine.

As we know from our previous lesson, electron orbits exist not as circular paths, but as regions in space around a nucleus in which an electron is allowed to exist. As two atoms get close to each other, these regions can overlap and interact with each other. This interaction can form a new region for the electrons to exist in. It is this new region which forms a bond.

An electron is bound to a nucleus in an atom by electrostatic attraction. If it drifts too far, the attraction between the positively charged nucleus and the negatively charged electron wanes, and the electron "falls" back into the "well" the nucleus creates, and the electron remains bound. When picturing one atom existing in isolation, this is easy to see. What happens if two atoms drift close to each other?

Imagine two hydrogen atoms.[1] Imagine that they have drifted close enough together that the distance between them is comparable to the volume of space that is occupied by the regions the electrons and protons exist in anyway. Now consider the shape of the regions that the electrons exist in.

We know from our previous lesson that the shape of the region an electron is likely to be in is defined in part by positively charged particles in the vicinity. Because electrons are attracted to protons, the regions in which they exist become biased, making it more likely to find an electron near any neighbouring protons than in an empty region of space away from the protons.

In the case of our two hydrogen atoms, we have two protons and two electrons. Each electron is attracted to both protons, and repelled by the other electron. With our current model of the atom, we see no particular reason for these atoms to bond together. Yes, a given electron will be attracted to the other nucleus, but it will be equally repelled by the other electron. They need a reason to connect, and some attractive force to hold them together. In other words, our picture of the atom is *still* incomplete.

8.4 Electricity and Magnetism

Throughout these lessons, the electric and magnetic forces have often been treated as a single entity. They are certainly related, but at this point, the interplay between these two factors needs to be made more clear.

Electrostatic attraction is relatively well known and understood. By the end of grade school, most public school students can tell you that like charges repel and opposite charges attract, and have rubbed balloons on heads and stuck them to walls at least once. This simple model is enough to discuss interactions when nothing is moving. When the electrically charged objects move, we need more.

Hans Christian Oersted was a sloppy enough experimentalist to become the first to discover the connection between electricity and magnetism. Experimentalists

---

[1] Hydrogen is the simplest atom to picture; it has one electron, one proton, and usually zero neutrons.

are taught to keep their experimental areas clear of anything which is not involved in the current experimental apparatus, simply to ensure that whatever is cluttered around did not impact their results. Oersted did not follow this practise consistently, and left a compass out on a counter-top when he went to work on an experiment involving circuits. He noticed that turning on the electric circuit caused his compass needle to move. Closer experimentation revealed that every moving charge, be it part of an electric current or a single particle moving through space, produced a magnetic field. This provides the link we need to turn our atoms into a molecule.

8.5 Nature's Smallest Bar Magnets

Every electron is a charged particle. Protons and neutrons are both made of quarks,[2] and those quarks have electric charges.[3] Could they produce magnetic fields? Experiment says they do. Every electron, proton and neutron behaves like a tiny bar magnet, and that behaviour makes all the difference in the world.

The bar magnet nature of electrons, protons and neutrons also shapes electron orbits. The interplay between these factors determines the shapes the allowed regions take, and they get quite bizarre. In some cases, they are simple spherical regions. They are nothing like planetary orbits, but they are the next most intuitive step.[4] It is when we start adding more than one electron that they get strange. They begin to show a preference for forming along the directions of the bar magnets themselves. They form teardrop shapes, donut shapes, ellipsoid shapes, jellybean shapes and more. Because the electrons orbit in alignment with the direction of their bar magnets, the molecules they form also align themselves with their internal bar magnets in many cases. When this happens consistently enough in a material, the atoms form something we see as a magnet in the macroscopic world.

Let us return to our pair of hydrogen atoms. With each particle acting like a bar magnet, the net attractive force on them is not zero. That extra little magnetic attraction, although much weaker than either individual electrostatic force, is just enough to nudge the electron orbits together and form a bond. Each electron no longer exists in a single "well" created by the nucleus it was originally bonded to, but now exists in a W-shaped well, in which it can travel between the two nuclei with relative ease. The electrostatic repulsion still keeps the electrons a fair distance away from each other, but that gap is tempered by the magnetic attraction. A bond forms, and the two hydrogen atoms form the simplest molecule in nature. This type of bond is known as a covalent bond.

---

[2] Section 2.3.2 on page 9 revealed that these particles have internal structure, and are made of other particles. These particles are called quarks.

[3] The total charge on the neutron is zero, but the charges on the particles it is made of are not zero themselves.

[4] At least, they are as intuitive as anything can be when you are dealing with quantum physics.

8.6 The Pauli Exclusion Principle

Most molecules are far more complicated than the simple hydrogen molecule. With our current picture, a bond formed between two carbon atoms (which have 6 electrons each) is much harder to predict. Is it a single, indistinguishable mishmash of 12 electrons? Do they somehow pair off to form the bonds, instead? If so, why not just pair off with other electrons that are already in the atom? There are a number of ways these particles can arrange themselves with our model. However, the arrangements in reality are actually quite well defined and restrictive. No atom easily forms more than four bonds, and most form fewer than four. Some atoms, for elements known as the noble gases, do not easily form even a single bond. Why is that the case? Why do different atoms have such wildly varying bond forming behaviours?

Wolfgang Pauli proposed a solution, referred to as the Pauli Exclusion Principle, which not only solved all of these problems, but allowed scientists to rearrange the periodic table entirely.[5] Pauli examined the mathematics describing electron orbits, and proposed an idea that fit experimental data perfectly: what if any given electron orbit could only be occupied by one electron at a time? When you know that electrons have zero volume, this seems an unnecessary feature, as multiple electrons could fit in the same allowed region as long as they didn't outnumber protons in the nucleus to the point that they drive the atom apart. Pauli proposed it out of necessity: at this stage, it needs to be driven into the theoretical model with a sledgehammer to explain experimental observations, but is not yet motivated by a theoretical need. It is required to form the particular types of complex molecules we see, but we do not yet see a need for these particular types of molecules to be favoured over the mishmash options mentioned earlier. This will be explored further in our final chapter. For the rest of this chapter, the implications of the idea will be used to answer most of our unanswered questions.

8.7 Insulators and Covalent Bonds

With the Pauli Exclusion Principle in place, the observed electron bonding options amongst electrically insulating materials makes complete sense. Each atom has certain orbits that electrons are allowed to be in, and once there is an electron in an orbit, that orbit cannot be occupied by another electron. More importantly, these orbits come in groups. The regions in which our electrons are allowed to exist are then restricted to empty orbits. The first group of orbits includes two possibilities for electrons to fill. This is why the first row of the periodic table has only two elements. The next group has eight orbits. This is why the second row of the periodic table has eight elements, and why carbon never easily forms more than

---

[5]When Dmitri Mendeleev created the first effective periodic table, it was arranged like a checker board, with empty spaces in half the squares and rules for moving through the table vertically, horizontally, and along both diagonal directions. The modern table is far more compact and effective, but Mendeleev's was good enough to predict several elements before they were discovered in nature.

8.8. INSULATORS AND IONIC BONDS

four bonds: the atom's six electrons fill the first group (of two) completely, and only fill half of the next eight.[6] When that carbon atom comes near another carbon atom, the electrons of the two atoms can expand their orbits across both atoms, filling in the same empty orbit for both atoms. Due to the way the magnetic interactions play out, the orbits within a group always come in pairs. This is why two electrons are involved in each such bond: the bonds form across the two unpaired electron orbits. These bonds also have distinct lengths due to the geometry of the orbits;[7] this is why materials formed this way have volume. They do not have volume because their constituent atoms have volume, but because the bonds that hold the atoms together have volume.

There is no reason to restrict this model to two atoms. Electrons can pair off between atoms, and in cases such as carbon (which can create four bonds in pairs like these) they form these bonds with four different atoms if at all possible.[8] This is how the molecules of electric insulators form. The atoms are bound together with relative security, but the electrons are bound to the two nuclei whose orbits they complete. The electrons are not free to travel throughout the entire material.

8.8 Insulators and Ionic Bonds

In some materials, bonds are formed without the use of covalent bonds. A covalent bond is, in effect, a way for an atom to "pretend" that it has more electrons than it really has. One of its own electrons starts to spend half of its time near another atom, while one of the other atom's electrons spends half its time near the first atom. This farce might be easy to pull off if the entire group of orbits is only short a few electrons, but what happens when it takes a relatively large number of electrons to fill the orbits? If we look at sodium instead of carbon, we find an atom with its first group of orbits filled to capacity and a single electron in its next group of eight available orbits. Filling seven orbits with "half time" electrons will not be particularly convincing if the atom is trying to pretend all orbits are full. Instead, sodium takes the easy way out: it dumps off the electron entirely.

When the electrons pair off in their orbits, the paired electrons feel a strong magnetic bond. If you put sodium and chlorine in close proximity, you get a violent chemical reaction. This is because the third electron from sodium bonds with the chlorine in its empty orbit so strongly that it is ripped from the sodium entirely. When large quantities of sodium and chlorine are brought together, this process happens on a massive scale. Electrons are violently ejected and recaptured by

---

[6]The groups also have subgroups, and the subgroups fill in sequence. This is why so few atoms have more than four bonds at a time; no subgroup holds more than eight electrons. The reason the bond limit is half the number of orbits in a subgroup will be covered in the next section.

[7]These orbits have volume because they, like the orbits around single atoms, are determined in part by the wavelengths of the electrons themselves.

[8]One atom can certainly form up to three bonds with a single neighbour. However, doing so means the bonded electrons get "bunched up" along the geometric line that can be drawn between the two nuclei, and the electrostatic repulsion gets harder and harder to overcome. It takes a lot less energy to form the bonds in four different, widespread directions.

neighbouring atoms, releasing significant heat energy in the process.[9]

When sodium and chlorine combine in this fashion, we get regular table salt. However, the picture seems incomplete. Salt is solid, and yet the above description does not appear to have any rigid bond connecting two neighbouring atoms. How are the components of salt arranged in a way that gives it volume?

If we picture just two atoms making an electron exchange of this type, we would expect them to collapse in on each other. After all, the sodium now has more protons than electrons, making it positively charged, while the chlorine is now negatively charged. They are made of zero volume particles and do not shared a rigid covalent bond, so there is no reason for them to stay apart from one another. Part of the problem is that we are picturing only two atoms, and not a large number of them.

First, picture two sodium atoms and one chlorine atom of each type. If the electrons have already been exchanged, then the sodiums will both be attracted to the chlorine, and vice versa. However, the sodiums will also be repelled by each other. There is only one stable arrangement for the chlorine: the three need to be equally spaced along a line with the chlorine in the middle. While the chlorine is in the middle, it feels no net force, as it is pulled equally in opposite directions. If it starts to drift off of the line in space that the sodiums are on, then the net attraction pulls it back to that line. The sodiums repel each other, but not as strongly as the chlorine attracts them. The chlorine isn't going anywhere, but the sodiums would seem to collapse in on it.

We are one step closer to building solid salt. We now have a chlorine pinned in place. In fact, if we extend this picture in all three dimensions we exist in, we can pin it down even further: if the chlorine is in the middle of a three dimensional cross, then that chlorine is pinned in all directions. The sodiums forming the cross are now much less stable, though: there are now six of them around, all repelling each other. We can counteract that by adding more chlorines behind them along the lines of the cross, and even surrounding them with more chlorines around them in the other two dimensions. We end up with a picture of the true structure of salt: sodiums and chlorines alternate in all three directions, forming a fairly regular pattern. They can jitter back and forth on the spot a bit, but for the most part, they stay where they have been placed. In the real world, it is not a perfect pattern. To start, the particles on the edges are always more weakly bound to the structure than the others, which is one reason salt (and other materials) can break. It is also rare to find 100% pure sodium and chlorine, so we often have other materials in the middle mucking up our crystals. Finally, random motion from heat also comes into play while the crystals form, causing deviations in the crystal that are surrounded by properly made crystals. These deviations can be hard to repair without breaking another part of the crystal structure, due to the tightly packed nature of them. Still, the constant electrostatic tug of wars between neighbouring atoms keeps the atoms

---

[9] If it takes 15 units of energy to hold an electron in place in its atom, but only 10 units of energy once it has bonded, 5 units of energy will be replaced in the process. These reactions happen naturally, such as combustion, and are referred to as exothermic reactions. Endothermic reactions are when you give energy back to the system and force the bonds to reverse themselves. For example, the extra electron in chlorine has enough energy to break free and get recaptured by a nearby sodium.

8.9. CONDUCTORS

bound together closely enough to seem solid, while still far enough apart to form a solid with volume.

8.9 Conductors

Some materials conduct electricity very well. The electrons within them are free to move throughout the structure, but neither of the pictures we have formed yet can explain this behaviour.

If you will recall, in covalent bonds the electrons are tightly bound to two nuclei, and it is these rigid bonds that hold the material together. The atoms require less energy to hang on to electrons when a group of orbits are either completely filled or completely empty. The covalent bonds typically form in materials which have groups of orbits that are nearly full. Metallic bonds, on the other hand, exist when the groups of orbits in the atoms are nearly empty.[10]

When two metallic atoms get close to one another, they form a bonds much like covalent bonds using electrons from their partially full group of orbits. Things change when a third atom is introduced. With these atoms, it takes less energy to expel electrons than to capture them, and that pattern continues. The bonds joining the first two atoms expand to include the third. As each new atom joins the structure, a new bond forms throughout the entire structure for each electron in the incomplete group of orbits. Eventually, there are so many bonds involved that there are available orbits with only slightly different energies, leaving the electrons to jump from orbit to orbit and move freely through the crystal. They do so constantly, but randomly, so there is no overall current in a typical metal sample. However, when a voltage is applied across the material, then there is a direction in which electrons can flow which leaves them with less energy.

In many ways, electrons behave like fish in a fish tank. Fish have limited mental capacity; in the absence of food and predators, they will naturally spread out, and move virtually at random throughout the fish tank. If you take those same fish and put them in a slow moving stream, also without food or predators, they will still choose their own directions at random, but will drift along with the current. Electric currents move much the same way; the voltage creates an electric current just as gravity causes a water current. The difference is that an electric circuit can be downstream all the way, unlike a water current.[11] The atoms are still bound by bonds with volume, and the material they form will therefore also have volume.

8.9.1 Electrical Resistivity

The picture of electrical resistance covered in most public school systems is the Debye model, named after its developer. In this early model, electrons experienced

---

[10]The noble gases do not readily form bonds because their groups of orbits are naturally full.

[11]Exceptions to this rule for water current were once reported by Rick Marshall, but this claim has been brought into doubt by the facts that he claimed he and his children also found living dinosaurs and hostile, intelligent, reptilian Sleestak in the same location, and that he cannot produce a convincing reason for abandoning his children in such an environment.

electrical resistance when they collided with other nuclei as they moved through the material. With nuclei made out of zero volume particles and orbits that cover the entire material, this picture is called into question. Even more damning, when experiments were performed to test the theory, they proved the theory was wrong.[12] It is still taught because it is a simple model to picture, so students can walk away thinking they understand reality, but that does not make the model correct. As Einstein once said, "things should be made as simple as possible, but *not* simpler."

More advanced calculations indicate that the actual truth is related to the imperfections and defects in the crystal already mentioned. In a perfectly formed solid, electrical resistance is zero. The fewer irregularities there are in a crystal, then the better it conducts. Perfectly formed solids do not exist in nature at any attainable temperature. However, at different temperatures, it is possible that many of these irregularities smooth themselves out as the atoms jitter around more or less, creating a path through the material which avoids the defects. Electrons can then travel that path, as long as there are enough orbits available passing through it for all of the electrons to fit. In superconducting materials, the electrons find ways to partner up and circumvent the Pauli Exclusion Principle[13] so that they can all take this path, and the measurable resistance to current flow drops to zero.

8.9.2 Photoelectric Conductors and Semiconductors

Photoelectric materials are a form of semiconductor, meaning they are neither pure conductors nor pure insulators.[14] In these materials, there are multiple types of bonds available.

Most electrons find themselves in low energy covalent bonds when they are in semiconducting materials. It is possible, however, for these particles to move into orbits formed by metallic bonds, if they have enough energy. If enough voltage is applied across the material, a weak current flows, as some of these electrons are driven into the metallic bond orbits that conduct well, but not many electrons get there. For photoelectric materials, that energy comes not from a battery or voltage source, but from the incoming light. The electrons in the covalent bonds capture light. The manner in which they do this not only reverses the direction of that particle's bar magnet,15 but gives it enough energy to jump up to the higher energy conductive orbits. The gap in energies between the covalent bonds and the conductive bands is the material's work function. This is why we have a well defined work function that allows continuous energies while conducting, as we have been wondering since chapter five.

---

[12]There are degrees of wrong. When experiments started proving Newton wrong, it was by a small degree, indicating that Newton's theories needed to be tweaked in some way, but were a useful point to build off of. It was akin to having theory predict that a certain measurement would be between 999,999 and 1,000,001, but experiments measure 1,000,005. The number is not very wrong, but it is outside the acceptable range. The Debye model, on the other hand, is not even close. If Debye theory predicts an experimental value between 999,999 and 1,000,001, then the actual experiment might measure 50.

[13]We will see how this works in the next chapter.

[14]In actuality, very few materials are pure in one sense or the other, but these materials fall very close to the middle.

Chapter 9

Like a Record, Baby

9.1 Unanswered Question

1. Why do particles act like bar magnets?

2. Why does Pauli's Exclusion Principle exist in nature? If these particles have zero volume, why can't two electrons share the same orbit?

9.2 Angular Momentum

Before we get back to answering our questions about quantum mechanics, we need to lay out an important part of classical mechanics.

Classical mechanics involves a quantity known as "angular momentum," which is conserved. Most people will have an understanding of what those two words mean in isolation, but the combination is something of a mystery.

Imagine a figure skater on an ice rink. This skater starts to spin on the spot, with her arms out.[1] She then folds her arms in, and as a result, starts to spin faster. When she stretches her arms out again, she slows down again. No outside forces are acting on her, and she isn't pushing along the ice; she can do the same thing whether on the ice or in the air. How does that work?

This works because the quantity known as "angular momentum" is conserved. This is a combination of two factors:

1. Position: where an object is, and

2. Momentum: how quickly an object with given inertia is moving

The rest of this section gets into heavy mathematical logic, although the actual algebra is minimal. If you are satisfied knowing only that angular momentum depends on position and momentum, and that if either position or momentum is zero

---

[1]The logic holds true for male skaters, too.

then the angular momentum is zero, then feel free to skip ahead to the next section if the going gets rough.

The way position and momentum combine to form angular momentum depends on both the magnitude of these quantities, and their directions, and is most obvious when the motion is circular. If the object in motion is tied to a pole, or a spinning ice skater, then it is forced to move in a circle. Let us choose the skater's right hand as the object of interest. Assume that the ice skater is rotating around a line that goes through her exactly down her middle. Assume also that, when viewed from a camera directly above her head, she is spinning counter clockwise.

If her arm is completely outstretched, then the magnitude of its position is half her shoulder width plus the length of her arm, as we measure from the skater's centre, which she rotates around. The direction from her centre to her hand is directly along her arm.

The momentum of her hand will have a given magnitude according to how rapidly she is currently spinning. The direction is the tricky part. Picture the skater in a moment when she is facing due North. At this moment, her right hand is extended due East. Because we decided she is spinning counterclockwise when viewed from above, her right hand is travelling toward the North at this instant. In fact, the position of her hand is always to her right, and the direction of travel is always the direction she is facing, perpendicular to her outstretched arm.

The angular momentum when position and momentum are perpendicular is merely the product of the two quantities. So, if the position of her hand is 1 m to her right,[2] and the momentum of her hand is 1 in the appropriate units, then her angular momentum is also 1.[3]

Now imagine she pulls her arms in, and the magnitude of the position of her right hand becomes a quarter of what it used to be. Angular momentum is subject to conservation: if we cut the position by four, the momentum must be multiplied by four. This is why the skater speeds up in her spin when she pulls her arms in; the product of the positions of her hands, arms, sides, and other body parts with their respective momenta must be constant. If her body parts get closer to the centre, the positions get smaller in magnitude, and the skater speeds up.

9.3 Spin

Any time a charged particle moves, it produces a magnetic field. All magnetic fields carry angular momentum. The reasons for these phenomena will not be dealt with here.[4]

---

[2] She's a gangly one, our figure skater.

[3] The angular momentum has a direction, too, although that won't be terribly important for our purposes. In this case, the direction is up. Take your right hand, open it flat, point your thumb in the direction of her hand's position, and point your fingers in the direction of her hand's motion: your palm now points in the direction of the angular momentum.

[4] Before Einstein, the first phenomenon was jammed into the theory quite forcefully because the behaviour was noted in the lab. Once that had been established, the second followed, not by logic, but "because the math says so." After Einstein, both can be explained through the theory of relativity with charged particles, but that is well beyond the scope of this series. Perhaps

What we have with electrons, protons, neutrons, and their component particles are electrically charged particles that exist in regions of probability. When these regions overlap, they interfere with each other, and the manner in which they interfere and interact depends on the fact that each of these particles acts like a little bar magnet.

These particles act like bar magnets because they carry angular momentum. We call this angular momentum spin. Even more surprising, this angular momentum is a fundamental property of the particles. Every electron carries exactly the same amount of angular momentum, which is the same amount carried by a proton, the same by a neutron, and the same by the muons, taus and neutrinos mentioned several lessons back. Every photon carries the same angular momentum as every other photon, although the angular momentum of a photon is different from that of electrons, protons, neutrons and the rest. Other particles, such as the W, Z, graviton and gluon particles also carry angular momentum, or spin.

What are these particles rotating around? We don't know. Are they even rotating at all? We don't know. If they are rotating, we would need to determine what they are rotating around, and where the momentum came from in the first place. We would also need to explain why every electron carries angular momentum *identical* to that of every other electron, regardless of the situation. If we explore the behaviours and properties that particles have because of this spin, then perhaps we can sort out what spin really is.

9.4 Fermions and Bosons

The spin particles have come in two different varieties, which lead to two completely different sets of behaviour. We can, therefore, classify all particles by these behaviours. The categories are:

- **Fermions** - These particles are named after Enrico Fermi, who (with the help of Paul Dirac) was able to describe their behaviour. The fundamental point of the behaviour is that these particles obey the Pauli Exclusion Principle, as described last lesson. Electrons, muons, taus, neutrinos, protons, neutrons, and the quarks protons and neutrons are made out of fall into this category.

- **Bosons** - These particles are named after Satyendra Nath Bose, who (with the help of Albert Einstein) was able to describe their behaviour. They are not subject to the Pauli Exclusion Principle, and can coexist with others of their kind without a problem. Photons, gluons, W, Z, gravitons, and any other force carrying particles fall into this category.

These are two fundamentally different behaviours. Now, the problem arises when considering the differences between them. Fermions and bosons are all zero volume particles. Why would they possibly be limited in their ability to share space? The thought makes sense if the particles have volume, but they do not.

---

relativity will appear in some other year's summer school curriculum, should demand warrant another series.[5]

We do know this: the behaviour in question is a fundamental part of the universe we live in. If the Pauli Exclusion Principle didn't apply, it would be virtually impossible to form molecules with anything but hydrogen, as the electrons could all pile into a single orbit. Clearly, some property of these particles keeps them distinct and unable to share the same space. It is easy to imagine that the interactions of the fields themselves are what keeps them from sharing. However, that leads to another problem.

Bosons are made up of fields, just as fermions are. Yet, bosons are not subject to the Pauli Exclusion Principle. If they were, they would be unable to mediate forces as they do, because the number of virtual particles that could be exchanged would be limited. If they were limited, we would not see forces as the continuous actions they seem to be in the macroscopic world, and they would not be stable enough to hold our observable world together. In short, the universe we live in needs bosons.

Even more strange is the way they combine. In day to day life, we can put puzzle pieces together to make a picture, but we cannot put pictures together to make puzzle pieces. This is similar in the cases of fermions and bosons. An even number of fermions can combine to make one large boson, but no quantity of bosons can combine to make a fermion.

Again, this is strange, but necessary for what we see. Superconductivity works this way; the conducting electrons join up in pairs, and these pairs act like bosons. They can then all share a single, defect free orbit allowing them to traverse the entire solid with no electrical resistance at all.

Why do we need an even number of fermions to make a boson? It is nothing more than adding fractions. Every spin for a particle can be represented as a multiple of a constant called $\hbar$.[6] Bosons are whole multiples of this number, whether that multiple is 0, $\hbar$, $2\hbar$, $3\hbar$, etc. Fermions, on the other hand, are half multiples of $\hbar$, and can be $\frac{\hbar}{2}$, $\frac{3\hbar}{2}$, $\frac{5\hbar}{2}$, $\frac{7\hbar}{2}$, etc. When two or more particles bond themselves together, these spins can add or subtract (according to the bonds). If you combine an even number of fermions, adding or subtracting their spins gives you an even multiple of $\frac{\hbar}{2}$, which can then be reduced to a whole number multiple of $\hbar$. If you have an odd number of fermions, you are back to having an odd multiple of $\frac{\hbar}{2}$, and the combination is a fermion again. However, no amount of manipulation of boson spins through addition or subtraction will make fractions appear; bosons can only combine to form bosons.

9.5 Spin and Behaviour

This leaves a single, burning question:[7] why and how can a quantity like angular momentum determine whether or not particles can share orbits?

When Wolfgang Pauli first presented his reasoning for why the spin and behaviour of particles should be linked, he essentially wrote a seven page paper that

---

[6]Do not worry about the notation of this constant: it is, in essence, a really small number with units of angular momentum.

[7]Actually, there are still a LOT of burning questions in quantum physics, but this is the last one the author is going to point out in this series.

said "because the math says so." Many researchers were hoping for a conceptual reason for this connection. After all, math should not dictate a scientific theory. It can point us in directions to explore, and it can make it possible to use quantifiable experiments to verify theories; math is undoubtedly an indispensable asset to the scientist, and its critical importance cannot be underestimated. However, our mathematical models should be consistent representations of the physical models, and not the models themselves.

To that end, Raymond F. Streater and Arthur S. Wightman rewrote the argument in *PCT, Spin and Statistics, and All That*. It is a 199 page volume that says, in essence, that angular momentum and behaviour are connected "because the math says so."

Not satisfied with this, Ian Duck and E. C. G. Sudarshan endeavoured to expand this further. In *Pauli and the Spin-Statistics Theorem*, they strive to explain this connection at an elementary level. Their volume is 503 pages of "because the math says so."

Richard Feynman was famous for his ability to teach physics at any level to anyone. He maintained that, if you cannot explain a concept to a stranger in the time it takes to ride an elevator, then you do not truly understand that concept. Here is what Feynman, master teacher, had to say about this connection:

> Why is it that particles with half-integral spin are Fermi particles ... whereas particles with integral spin are Bose particles...? We apologize for the fact that we cannot give you an elementary explanation. An explanation has been worked out by Pauli for complicated arguments of quantum field theory and relativity. He has shown that the two must necessarily go together, but we have not been able to find a way of reproducing his arguments on an elementary level.... This probably means that we do not have a complete understanding of the fundamental principle involved....

In short, this is not a question that can be answered without mathematics at this time. The mathematical answer is called the "spin-statistics theorem," and can be found in the aforementioned resources. At the moment, the exact nature of spin and its connection(s) to the way particles behave is unknown. All we know for certain is that both types of particles exist, and a quantum physics theory that is consistent with Einstein's relativity requires them to exist. The specific reasons why are still a mystery.

9.6 Conclusion

Our knowledge and understanding of our world has increased dramatically over the last century or so, but we're still a long way from finished. Still, the concepts presented in these past nine lessons are those which have stood up to repeated and dedicated attempts to disprove them. That's not a complete guarantee that they are correct, but the power of science lies in its ability to place limits on how wrong we are. Any theory that eventually replaces these will show virtually identical behaviours under the laboratory environments that we have produced to date. However, traditionally speaking, every upgrade which expands our laboratory conditions has lead to wholly unexpected behaviour. With the LHC now operational, it's only a matter of time before the more tenuous ideas developed by man are either confirmed our outright replaced.

Bibliography

[1] Brian H Bransden and C J Joachim. *Introduction to Quantum Mechanics*. Longman Scientific and Technical, 1989.

[2] Ian Duck, Wolfgang Pauli, and ECG Sudarshan. *Pauli and the spin-statistics theorem*. World Scientific, 1997.

[3] Richard Clinton Fernow. *Introduction to experimental particle physics*. Cambridge university press, 1986.

[4] David Griffiths. *Introduction to Particle Physics*. Wiley, 1987.

[5] Claude Itzykson and Jean-Bernard Zuber. *Quantum field theory*. Courier Corporation, 2012.

[6] Aleksandr Iakovlevich Khinchin. *Mathematical foundations of quantum statistics*. Courier Corporation, 1998.

[7] Robert B Leighton and Matthew Sands. *The Feynman lectures on physics*. Addison-Wesley Boston, MA, USA, 1965.

[8] Jun John Sakurai and Eugene D Commins. *Modern quantum mechanics, revised edition*. American Association of Physics Teachers, 1995.

[9] Raymond Frederick Streater and Arthur S Wightman. *PCT, spin and statistics, and all that*. Vol. 52. Princeton University Press, 2000.

www.ingramcontent.com/pod-product-compliance
Lightning Source LLC
Chambersburg PA
CBHW070314220526
45465CB00004B/1859